THE SHEPHERD'S LIFE

These modern dispatches from an ancient landscape tell the story of deep-rooted attachment to place, describing a way of life that is little noticed and yet has profoundly shaped history. In evocative and lucid prose, James Rebanks takes us through a shepherd's year, offering a unique account of rural life and a fundamental connection with the land that most of us have lost. It is a story of working lives, the people around him, his childhood, his parents and grandparents, a people who exist and endure even as the world changes around them.

THE SHEPHERD'S LIFE

THE SHEPHERD'S LIFE

by

James Rebanks

Magna Large Print Books
Long Preston, North Yorkshire,
BD23 4ND, England.

British Library Cataloguing in Publication Data.

Rebanks, James
   The shepherd's life.

A catalogue record of this book is
available from the British Library

   ISBN   978-0-7505-4196-1

[Library stamp, handwritten:]

First published in Great Britain in 2015 by Allen Lane
who is part of the Penguin Random House Group of Companies

Copyright © James Rebanks, 2015

Cover illustration © mirrorpix

The moral right of the author has been asserted

Published in Large Print 2016 by arrangement with
Penguin Books Ltd.

Magna Large Print is an imprint of Library Magna Books Ltd.

Printed and bound in Great Britain by
T.J. (International) Ltd., Cornwall, PL28 8RW

Grateful acknowledgement is given to the following for permission to reproduce copyright material:

p.74 – from *The Poetry of Robert Frost* by Robert Frost, published by Jonathan Cape, reproduced by permission of The Random House Group Ltd; p.227 – from *Young and Old* by R.S Thomas, published by Chatto and Windus, reproduced by permission of The Random House Group Ltd; p.282–283 – 'The Past' by Kath Walker (Oodgeroo Noonuccal) from *The Dawn is at Hand,* published by Marion Boyars, 1992 (London & New York)

Dedicated to the memory of my
grandfather, W. H. Rebanks,
and with respect to my father,
T. W Rebanks

Towards the head of these Dales was found a perfect Republic of Shepherds and Agriculturalists, among whom the plough of each man was confined to the maintenance of his own family, or to the occasional accommodation of his neighbour. Two or three cows furnished each family with milk and cheese. The chapel was the only edifice that presided over these dwellings, the supreme head of this pure Commonwealth; the members of which existed in the midst of a powerful empire, like an ideal society or an organized community, whose constitution had been imposed and regulated by the mountains which protected it. Neither high-born nobleman, knight, nor esquire was here; but many of these humble sons of the hills had a consciousness that the land, which they walked over and tilled, had for more than five hundred years been possessed by men of their name and blood...

William Wordsworth,
A Guide Through the District of the Lakes in the North of England (1810)

Contents

Hefted   17

Summer   28

Autumn   139

Winter   227

Spring   282

Acknowledgements

Hefted

HEFT

Noun: 1) *(Northern England)* A piece of upland pasture to which a farm animal has become hefted.
2) An animal that has become hefted thus.

Verb: *Trans. (Northern England and Scotland) of a farm animal, especially a flock of sheep:* To become accustomed and attached to an area of upland pasture.

Adj.: Hefted: describing livestock that has become thus attached.
(Etymology: from the Old Norse hefð, meaning 'tradition')

I realized we were different, really different, on a rainy morning in 1987. I was sitting in an assembly in the 1960s-style shoddily built concrete comprehensive in our local town. I was thirteen or so years old, sitting surrounded by a mass of other academic non-achievers, listening to an old battle-weary teacher lecturing us on how we should aim to be more than just farm workers, joiners, brickies, electricians and hairdressers. It felt like a sermon she'd delivered many times before. It was a waste of time and she knew it. We were firmly set, like our fathers and grandfathers, mothers and grandmothers before us, on being what we were, and had always been. Plenty of us were bright enough, but we had no intention of displaying it at school. It would have been dangerous.

There was an abyss of understanding between that teacher and us. The kids who gave a damn had departed the year before to our local grammar school, leaving the 'losers' to fester away over the next three years in a place no one wanted to be. The result was something akin to a guerrilla war between largely disillusioned teachers and some of the most bored and aggressive kids imaginable. We played a 'game' as a class where the object was to smash the greatest value of school equipment in one lesson and pass it off as an 'accident'.

I was good at that kind of thing.

The floor was littered with broken microscopes, biological specimens, crippled stools and torn books. A long dead frog pickled in formaldehyde lay sprawled on the floor doing breaststroke. The gas taps were burning like an oil rig and a window was cracked. The teacher stared at us with tears streaming down her face – destroyed – as a lab technician tried to restore order. One maths lesson was improved for me by a fist-fight between a pupil and the teacher, before the lad ran for it down the stairs and across the muddy playing fields, only to be knocked down by the teacher before he escaped into town. We cheered as if it were a great tackle in a game of rugby. From time to time, someone would try (incompetently) to burn the school down. One boy who we bullied killed himself a few years later in his car. It was like being locked in a Ken Loach movie: if some skinny kid had turned up with a kestrel, no one would have been surprised.

On another occasion, I argued with our dumbfounded headmaster that school was really a prison and 'an infringement of my human rights'. He looked at me strangely, and said, 'But what would you do at home?' As if this was an impossible question to answer. 'I'd work on the farm,' I answered, equally amazed that he couldn't see how simple this was. He shrugged his shoulders hopelessly, told me to stop being ridiculous and go away. When people got into serious trouble, he sent them home. So I thought about putting a brick through his window, but didn't dare.

So in that assembly in 1987, I was day-dreaming through the windows into the rain, wondering what the men on our farm were doing, and what I should have been doing, when I realized the assembly was about the valleys of the Lake District, where my grandfather and father farmed. So I switched on. After a few minutes of listening, I realized this bloody teacher woman thought we were too stupid and unimaginative to 'do anything with our lives'. She was taunting us to rise above ourselves. We were too dumb to want to leave this area with its dirty dead-end jobs and its narrow-minded provincial ways. There was nothing here for us, we should open our eyes and see it. In her eyes, to want to leave school early and go and work with sheep was to be more or less an idiot.

The idea that we, our fathers and mothers, might be proud, hard-working and intelligent people doing something worthwhile, or even admirable, seemed to be beyond her. For a woman who saw success as being demonstrated through education, ambition, adventure and conspicuous professional achievement, we must have seemed a poor sample. I don't think anyone ever mentioned 'university' in this school; no one wanted to go anyway – people that went away ceased to belong; they changed and could never really come back, we knew that in our bones. Schooling was a 'way out', but we didn't want it, and we'd made our choice. Later I would understand that modern industrial communities are obsessed with the importance of 'going somewhere' and 'doing something with your life'. The implication is an idea I have come to hate, that staying local and doing physical work doesn't count for much.

I listened, getting more and more aggravated, as I realized that curiously she knew, and claimed to love, our land. But she talked about it, and thought of it, in terms that were completely alien to my family and me. She loved a 'wild' landscape, full of mountains, lakes, leisure and adventure, lightly peopled with folk I had never met. The Lake District in her monologue was the playground for an itinerant band of climbers, poets, walkers and daydreamers ... people whom, unlike our parents, or us, had 'really done something'. Occasionally she would utter a name in a reverential tone and look in vain for us to respond with interest. One of the names was 'Alfred Wainwright', another was 'Chris Bonnington'; and she kept going on and on about someone called 'Wordsworth'.

I'd never heard of any of them. I don't think anyone in that hall, who wasn't a teacher, had.

Sitting in that assembly was the first time I'd encountered this (basically romantic) way of looking at our landscape. I realized then, with some shock, that the landscape I loved, we loved, where we had belonged for centuries, the place known as 'the Lake District', had a claim to ownership submitted by other people, based on principles I barely understood.

Later, I would read books and observe the 'other' Lake District, and begin to understand it better. I'd learn that until around 1750 no one from the outside world had paid this mountainous corner of north-west England much notice, or, when they had, they found it to be poor, unproductive, primitive, harsh, ugly and backward. I'd be annoyed to discover that no one from outside seems to have thought it was beautiful or a place to visit until then – and yet be fascinated to discover how in a few decades that had all changed. Roads, and later railways, were built, making it much easier to get here. And the Romantic and picturesque movements changed the way many people thought about mountains, lakes and rugged landscapes like ours. Our landscape suddenly became a major focus for writers and artists, particularly when the Napoleonic Wars stopped the early tourists from going to the Alps and forced them instead to discover the mountainous landscapes of Britain.

From the start, this obsession was – for visitors – a landscape of the imagination, an idealized landscape of the mind. It became a counterpoint to other things, such as the Industrial Revolution, which was born less than a hundred miles to the south, or a place that could be used to illustrate philosophies or ideologies. For many, it was, from its 'discovery', a place of escape, where the rugged landscape and nature would stimulate feelings and sentiments that other places could not. For many people, it exists to walk over, to look at, or climb, or paint, or write about, or simply dream about. It is a place many aspire to visit or live in.

But, above all, I would learn that our landscape changed the rest of the world. It is where the idea that all of us have a direct sense of 'ownership' (regardless of property rights) of some places or things because they are beautiful, or stimulating, or just special was first put into words. The Lake District Romantic poet William Wordsworth proposed in 1810 that it should be 'a sort of national property, in which every man has a right and interest who has an eye to perceive and a heart to enjoy'. Arguments were formulated here that now shape conservation around the world. Every protected landscape on earth, every National Trust property, every National Park, and every UNESCO World Heritage Site, has a little bit of those words in their DNA.

Above all, in the years after I left school and grew up, I learnt that we are not the only ones that love this place. It is, for better or worse, a scenic playground for the rest of Britain, and for countless other people from around the world. I simply have to travel over the fell to Ullswater to see the cars streaming past on the roads, or the crowds milling around the shore of the lake, to see what this means. There are good outcomes and less good ones. Today, 16 million people a year come here (to an area with 43,000 residents). They spend more than a billion pounds every year here. More than half the employment in the area is reliant upon tourism – and many of the farms depend upon it for their income by running B&Bs or other businesses. But in some valleys 60 to 70 per cent of the houses are second homes or holiday cottages, so that many local people cannot afford to live in their own communities. The locals speak begrudgingly of being 'outnumbered', and all of us know that we are in every way a tiny minority in this landscape. There are places where it doesn't feel like it's ours any more, as if the guests have taken over the guesthouse.

So that teacher's idea of the Lake District was created by an urbanized and increasingly industrialized society, over the past two hundred years. It was a dream of a place for a wider society that was full of people disconnected from the land.

That dream was never for us, the people who work this land. We were already here doing what we do.

I wanted to tell that teacher that she had it all wrong – tell her that she didn't really know this place or its people at all. These thoughts took years to become clear, but in a rough childish form I think they were there from the start. I also knew in a crude way that if books define places, then writing books was important, and that we needed books by us and about us. But in that assembly in 1987 I was dumb and thirteen, so I just made a farting noise on my hand. Everyone laughed. She finished and left the stage fuming.

If Wordsworth and friends 'invented' or 'discovered' the Lake District, it didn't touch our family until 1987 when I went home and started asking questions about what the teacher had said. From the start, this other story felt wrong. How come the story of our landscape wasn't about us? It seemed to me an imposition, a classic case of what I would later learn historians call 'cultural imperialism'.

What I didn't know was that Wordsworth believed that the community of shepherds and small farmers of the Lake District formed a political and social ideal of much wider significance and value. People here governed themselves, free of the aristocratic elites that dominated people's lives elsewhere, and in Wordsworth's eyes this provided a model for a good society. Wordsworth thought we mattered as a counterpoint to the commercial, urban and increasingly industrial England emerging elsewhere. It was an idealistic view even then, but the poet's Lake District was a place peopled with its own culture and history. He believed that with the growing wider appreciation of this landscape came a great responsibility for visitors to really understand the local culture, or else tourism would be a bludgeoning force erasing much that made this place special. He also recognized in these discarded lines from a draft of 'Michael, a Pastoral Poem' (written in 1800) that a shepherd's view of this place was different and of interest in its own right, a remarkably modern observation:

No doubt if you in terms direct had ask'd
Whether he lov'd the mountains, true it is
That with blunt repetition of your words
He might have stared at you, and said that they
Were frightful to behold, but had you then
Discours'd with him in some particular sort
Of his own business, and the goings on
Of earth and sky, then truly had you seen
That in his thoughts were obscurities,
Wonders and admirations, things that wrought
Not less than a religion in his heart.

But for a long time I knew none of this, and blamed Wordsworth for the failure to see us here and for making this a place of romantic wandering for other people.

We are all influenced, directly or indirectly, whether we are aware of it or not, by ideas and attitudes to the environment from cultural sources. My idea of this landscape is not from books, but from another source: it is an older idea, inherited from the people who came before me here.

What follows is partly an explanation of our work through the course of the year; partly a memoir of growing up in the 1970s, 1980s and 1990s and the people around me at that time, like my father and grandfather; and partly a retelling of the history of the Lake District – from the perspective of the people who live there, and have done for hundreds of years.

It is the story of a family and a farm, but it also tells a wider story about the people who get forgotten in the modern world. It is about how we need to open our eyes and see the forgotten people who live in our midst, whose lives are often deeply traditional and rooted in the distant past. If we want to understand the people in the foothills of Afghanistan, we may need to try and understand the people in the foothills of England first.

Summer

I've lived in the country for a lot of my life but I've never felt that I belonged... It is so strange ... I have never experienced such an atmosphere ... as exists here... I have to talk about it simply because it is so curious. It is the power which the children have to resist everybody and everything outside of the village... The village children ... are convinced that they have something which none of the newcomers can ever have, some kind of mysterious life which is so perfect that it is a waste of time to search for anything else.

Daphne Ellington, teacher, quoted in Ronald Blythe, *Akenfield* (1969)

There is no beginning, and there is no end. The sun rises, and falls, each day, and the seasons come and go. The days, months and years alternate through sunshine, rain, hail, wind, snow and frost. The leaves fall each autumn and burst forth again each spring. The earth spins through the vastness of space. The grass comes and goes with the warmth of the sun. The farms and the flocks endure, bigger than the life of a single person. We are born, live our working lives and die, passing like the oak leaves that blow across our land in the winter. We are each a tiny part of something enduring, something that feels solid, real and true. Our farming way of life has roots deeper than five thousand years into the soil of this landscape.

I was born in late July 1974, into a world that centred on an old man and his two farms. He was a proud farmer called William Hugh Rebanks, 'Hughie' to his mates, 'Granddad' to me. He had a rough whiskery face when you kissed him goodnight. He smelt of sheep and cattle, and only had one yellow tooth, but he could clean the meat off a lamb chop with it like a jackal.

He had three children. Two daughters, who had married good farmers, and my father. Dad was the youngest, the one who was to carry on his farm. I was his youngest grandson, but the only one with his name. From my first memories until his dying day, I thought the sun shone out of his backside. Even as a small child, I could see that he was the king of his own world, like a biblical patriarch. He doffed his cap to no man. No one told him what to do. He lived a modest life but was proud and free and independent, with a presence that said he belonged in this place in the world. My first memories are of him, and knowing I wanted to be just like him some day.

We live and work our small hill farm in the far north-west of England, in the Lake District. We farm in a valley called Matterdale, between the first two rounded fells (mountains) that emerge on your left as you travel west on the main road from Penrith. From the summit of the fell behind our house, you can see north across the silver glimmering of the distant Solway estuary to Scotland. There is a stolen moment each early summer when I climb that fell and sit with my sheepdogs and have half an hour to take the world in. To the east you can see the backbone of England, the Pennines, with the good farming land of the Eden Valley opening up below. I smile at the thought that the entire history of our family has played out in the fields and villages stretching away beneath that fell, between Lake District and Pennines, for at least six centuries, and probably longer. We shaped this landscape, and we were shaped by it in turn. My people lived, worked and died down there for countless generations. It is what it is because of them and people like them.

It is, above all, a peopled landscape. Every acre of it has been defined by the actions of men and women over the past ten thousand years. Even the mountains were riddled with mines and pocked with quarries, and the seemingly wild woodland behind us was once intensively harvested and coppiced. Almost everyone I am related to and care about lives within sight of that fell. When we call it 'our' landscape, we mean it as a physical and intellectual reality. There is nothing chosen about it. This landscape is our home and we rarely stray far from it, or endure anywhere else for long before returning. This may seem like a lack of imagination or adventure, but I don't care. I love this place; for me it is the beginning and the end of everything, and everywhere else feels like nowhere.

From that fell, I look out over a place crafted by largely forgotten working people. It is a unique man-made place, a landscape divided and defined by fields, walls, hedges, dykes, roads, becks, drains, barns, quarries, woods and lanes. I can see our fields and a hundred jobs that I should be doing instead of idling up on the fell. I see sheep climbing a wall into a hay meadow down below, and I know I have to stop messing about, daydreaming like a bloody poet or day-tripper and get some work done. To the west, I see the high fells of the Lake District, often covered for half the year in snow, and from the highest of those you can see the Irish Sea. To the south, the fells block my view, but somewhere beyond them is the rest of England. The Lake District is relatively small, being only about 800 square miles. So, if you looked down on our land from outer space, you would see we are on the eastern edge of a small cluster of mountain valleys. Our valley is small, even by the standards of the Lake District, a basin of enclosed land and meadows surrounded by fells, scattered with little farmsteads. I can drive through it from one end to the other in five minutes. I look across to my neighbours on the other side of the valley a mile away and can hear them gathering their sheep on the fell sides. The valley where we live and farm stretches beneath me like an old man's upturned cupped hands.

There is something about this landscape that people love. It would, in summer, seem to most people around the world to be exceptionally green and lush. It is a 'pastoral landscape' and 'temperate', a place of heavy rainfall and warm summers, an excellent place, in short, for growing grass in the summer. As writers have long noted, it is an intimate landscape on a human scale. Whitewashed farmhouses hug the fell sides just beneath the ancient common land of the fells. Other farmsteads dot the valley floor on higher ground, or riggs, that rise from the rushes of the sodden land in the valley bottom, including the one where my grandfather lived. We are one of maybe 300 farming families who sustain this landscape and its ancient way of life.

My grandfather was born in 1918 into a

fairly anonymous and unexceptional farming family. At that time they mostly lived and farmed down in the heart of the Eden Valley. The written records, for what they are worth, show that my grandfather belonged to an agricultural family struggling by from generation to generation, occasionally making it into the ranks of relatively established farmers, before sinking back into being tenants, or farm-workers, or in the workhouse, or worse. The written story peters out into an illegible sixteenth-century script of births, deaths and marriages, in church records belonging to little villages close to where their descendants still live and work. My grandfather is, quite simply, one of the great forgotten silent majority of people who lived, worked, loved and died without leaving much written trace that they were ever there. He was, and we his descendants remain, essentially nobodies as far as anyone else is concerned. But that's the point. Landscapes like ours were created by, and survive through, the efforts of nobodies. That's why I was so shocked to be given a 'dead, rich, white man' version of its history at school. This is a landscape of modest, hard-working people. The real history of our landscape should be the history of the nobodies.

The alarm clock vibrates on the bedside table. My hand swipes across and kills it:

4.30 a.m. I was only half-asleep anyway. The room is already half-lit with the coming dawn. I see my wife's shoulder, and her leg curled over the sheet, and my two-year-old son lying between us, where he came in during the night. I move quietly out of the room with a fistful of clothes. The sun will rise soon over the edge of the fell.

In the kitchen I swig at a carton of milk. I throw on my clothes robotically, half-awake. I have half an hour before we are meeting at the fell gate. We are going to gather the fell flock in for clipping (shearing). My mind is on a kind of checklist autopilot.

Right clothes: check.

Breakfast: check.

Sandwiches: check.

Boots: check.

As I get to the barn, my sheepdogs Floss and Tan jump, wriggle and make whining noises until I unchain them. They know we are going to the fell. I feed them so they'll have energy later when they need it. A shepherd on a fell without a good sheepdog, or dogs, is useless. The fell sheep are half-wild, can smell weakness, and would escape and create chaos without good sheepdogs. Men can't get to lots of places the dogs can – to the crags and rocky screes, to chase the ewes down. When I head out, Tan bolts for the barn door and jumps on the quad bike. Floss follows.

Sheepdogs fed and loaded: check.

Quad bike: check.

Fuel: check.

The swallows explode outwards from the barn door, disturbed by the dogs. They fledged a couple of days ago and whole families head out over my head to the fields where they hawk all day over the grass and thistles.

Fingers of pink and orange light are now creeping over the fell sides. Sunrise.

These are the hottest days of summer. As I go along the road, I feel the heat rising from the tarmac. Sun. Dust. Flies. Blue skies. It is too hot in the heat of the day for moving sheep, something we would scarcely have believed possible for the past eight or nine months of cold wet weather. By midday they will be panting, or hiding in the nooks and crannies for shade, and we will miss lots of them. It is too hot for sheepdogs as well. You can kill dogs working them too hard in the heat and humidity. So we intend to start early and do the work before the sun burns high in the sky.

I didn't know anything about gathering today until last night. I had been in the bath when the phone rang. My wife brought it in and I pretended I wasn't in the bath. It was my neighbour Alan, an older, well-respected farmer who has a lot of sheep on the fell and has done it much longer than me. He's the boss, the elder statesman, if you like, and he organizes the commoners to work together. Organizing fell farmers to do anything collectively is not easy, so I don't envy his job one bit. He doesn't waste words unnecessarily.

'We are gathering the fell tomorrow.'
'OK.'
'Meet at the fell gate at 5 a.m.'
'Right.'

Then he hangs up to call someone else.

I knew it was impending because of the date, and because it is time to clip the ewes, but it is a communal job that needs the right weather, and men to be free of other work to do it. So it's a bit like waiting for D-Day – you never know until the phone call, or shout from the road as he passes, to say 'It's on tomorrow'.

Gathering is ancient communal work that consists of everyone with rights to graze sheep on the unfenced common land working together with their sheepdogs to bring in the flocks from the fells. There are about ten different flocks of sheep on our fell, a vast unenclosed piece of moorland and mountain. Because there are no large predators, the sheep are left to graze alone, but are brought down several times a year for lambing, clipping and other key activities in the life of the flock. Beyond our common lie other unfenced areas of mountain land, other fells, farmed by other commoners, so in theory our sheep could wander right across the Lake District. But they don't because they know their place on the mountains. They are 'hefted' – taught their sense of belonging by their mothers as lambs – an unbroken chain of learning that goes back thousands of years. So the sheep can never be sold from the fell without breaking that ancient link. This is, they say, the greatest concentration of common land in Western Europe; and on it survives a kind of farming that is older than that which exists across much of the world today.

The fell land we are gathering today doesn't belong to us, it belongs to the National Trust. Other fells belong to other landowners, but we have an ancient legal right to graze a set number of sheep on it. Many of these mountainous areas of land were bought and given to the National Trust by wealthy benefactors like Beatrix Potter, who trusted them to protect the landscape and its unique way of life. The bequests often stressed that the fell flocks had to remain Herdwick sheep.

There are different kinds of ownership on one piece of land. The grazing rights on our fell are divided into something called 'stints' (a share of the common rights); and each stint you own, or rent, entities you to graze a certain number of sheep (six per stint on our fell). We buy and sell and rent stints so that older farmers can retire and their grazing rights and flocks can be taken forward by the next generation. The owner of the fell sometimes owns no stints and cannot therefore graze his own land unless there are surplus grazing rights. The rights to graze are held in common with our fellow commoners. 'Commoner' isn't a dirty word here; it is a thing to be proud of. It means you have rights to something of value, that you contribute to the management of the fells, and that you take part in our way of life as an equal with the other farmers. If you farm Herdwick or Swaledale sheep and they are hefted to the common grazing land on the fells, then you, by definition, often belong to an association of 'commoners'. This is all a strange hangover from a feudal past when we paid dues (including bearing arms) to the Lord of the Manor in return for the right to graze the poor mountain land. But no dues have been paid for a long time now. The aristocrats either disappeared or couldn't be bothered to contest our rights, because we are troublesome and stubborn when crossed. It was more effort than it was worth, so we, the peasants, won. We are a tiny part of an ancient farming system and way of life that somehow has survived in these mountains because of their historic poverty, relative isolation, and because it was protected from change by the early conservation movement.

My ewes and lambs have been up in the mountains for nearly eight weeks. They are Herdwick sheep, native to the Lake District fells and bred for centuries to suit this landscape, this climate and this way of farming. They have two functions: survive the winters and the tough times, and in the spring and summer months produce good lambs and rear them in the mountains so the flock is sustained with ewe lambs and the farms have a surplus of lambs to sell.

In the eight weeks since I brought them here, I have not seen many of them. They have looked after themselves on the abundant summer grass. Our shepherding culture includes periods when the sheep graze the fells away from our supervision. Only the ewes with twins stay down on the lower slopes on our own fenced land, called 'intakes' or 'allotments', because they need better nutrition to rear twins than the mountains offer. So I am anxious to see them again, keen to check that they are alive and well. Above all, I am interested to see how much my lambs have grown since I brought them up when they were just a month old in May. It is now the second week in July. Mist hangs in the hollows as I head across the high ground to the fell gate. It is already starting to burn off with the rising sun.

I reach the fell gate second. One shepherd always gets there first. I suspect he is an insomniac.

Fell gate on time: check.

Soon the fell gate is a meeting place for eight or ten men and women. An assorted pack of sheepdogs, and other willing mongrels, circle excitedly. Occasionally, there is a snarl-up. Everyone is in short sleeves, booted and in an array of sunhats that won't win any fashion prizes. Over shoulders are slung tatty old bait-bags, packed with sandwiches, pop and cake. On bad days we stare nervously at the skyline and the clouds hugging the fells. Sometimes we have to turn back if the clouds are too low, and return later. It is dangerous up there in bad weather. Snow makes it potentially lethal. But today there is only one worry: the heat. One of the shepherds is late, so everyone is impatient and frustrated. We stand and curse him.

'He is always late.'
'Can't get up, that bugger.'
'Let's go without him. He will catch up.'
'No, we better wait.'
'Oh, here he is.'

A quad bike races up the fell-side road. A slightly flustered shepherd mumbles his apologies. He has been gathering up some lambs down below that had escaped on to the road.

It doesn't matter. We need to get going. Move fast. The ewes and lambs are high up on the fells where the land meets the sky.

The oldest shepherd performs the function of a general on a battlefield. There is a bit in the movie *Zulu* when the natives' battle plan is described like the 'horns of a buffalo ... that come around like pincers and encircle you'. That's a bit like the way we gather our fell. It takes six or eight people and a dozen or more dogs, involves hours of walking (though is made a little quicker by a quad bike on the driveable bits) and requires everyone to work more or less as a team. As you pass over the fell, you try to use your judgement to carve through between the flocks of our common and the sheep of the next, by judging their 'smit marks' – the coloured paint marks that identify the sheep to specific farms. Anyone ignorant of the flocks and the marks and the lie of land can make a terrible mess and push sheep on to a neighbouring common and thus make unnecessary work for everyone. We stand and chat, but it's a serious business. We must do what we're told. No fucking around.

One of the most experienced shepherds, called Shoddy, is sent over the fell tops to clear out some distant crags, high up where the green meets the blue. The best men and dogs are sent to the hardest places. He will define the far end of the gather, and act like a blocker when the sheep try to flee away from us, tucking them back down at the far end.

Joe, a younger fell shepherd with good dogs, is sent to clear out a long deep ravine (we call them 'ghylls', carved out by the beck over many centuries) on the left-hand arm of the gather where our common meets the next one. A great dog can bring sheep carefully out of the crags, moving left or right or stopping on a sixpence at a whistled command. A young or poorly trained dog would just fail to get them down, or worse, scare them into danger on the scree or rock faces.

These are good fell shepherds with a pack of good dogs apiece. They disappear off, one on a quad bike, the other loping off across the heather.

Two or three of us are sent up the left-hand side of the fell, after Joe, to sweep out the sheep across the fell to the right, with one of us peeling off to hold them that way every half mile or so. Each of us has a landmark we are to hold at.

Each of us is responsible for not letting any sheep break back past us – easy with a good dog, impossible without one. Farming the fells is only possible because of the bond between men and sheepdogs.

I'm the last one on this arm of the gather.

I am to meet Shoddy at the far end. Wait at the Stones for the others, I'm told. Right.

The eldest shepherd takes a couple of men with him along a dusty old track to the right. He will form a break before the next common, pushing their sheep away and fetching ours back: he will form the right arm of the gather.

Men bawl to their dogs, who are excited and heading off after the wrong shepherds. We will meet them in a few hours at the far end, past the peat hags – raised peat bogs that rise up out of the sward, like green, or brown, islands slowly emerging out of the earth. They form a sea of raised mounds, some twenty or thirty feet across, others acres in size. They are carved apart by little gulleys and valleys worn by the water, form-ing dangerous cliffs of black peat the height of a man, or deeper, that you can tumble into. The sheep rub their backs on these peaty cliff faces, giving their fleeces a coal-black hue that tells us this is where they live. In the sheltered low ground between the peat hags, sheep can be lost from sight and the quad bike can be easily turned over, so you have to pay attention to navigate through the bogs and ensure the flock is cleared out of them and pushed by the dogs away homewards. Beyond them, we meet at Wolf Crag and form a kind of noose, with all of the fell encircled and the sheep heading in the right direction for home.

After the noise at the fell gate gathering, it quickly becomes a quieter and lonelier day's work. Most of it is spent far from other people, working with them, but far beyond talking distance. It is a day to work with the dogs. A fell dog is a special thing, tough as old boots, smart and capable of working semi-independently a long way across the mountain. I'm a lucky man to have two fine 'field' sheepdogs – Border collies. There isn't much they can't do in the valley bottom. They'll creep and crawl, and dart every which way, and hold sheep spellbound with a look. They are my pride and joy, but they are not great fell dogs (not yet anyway). That's a totally different thing altogether. Fell dogs are their own type: they need to be strong and smart, and less about eye and more about following instruction or using their wits when beyond command.

As we head across the fell, we see some ewes that should be on our common beyond a deep ghyll on the mountainside opposite. I fear they are too far away to get them today. They will, I assume, come in with the neighbouring common and we will collect them later. But Joe, who is cleaning out that ghyll, has sent his dogs to get them. From where he is, he can scarcely see the sheep they are so far away. He is further away than we are. The dog lurches back, onwards, up and up, climbing higher and higher towards the distant skyline. A whistle or two reassures it that it should keep going for sheep it cannot see yet because of the lie of the land. Then the dog sees the sheep it has been sent for, and knows what to do. It circles behind them and pushes them out of the crags. They twist and turn ever downwards and back towards us, then disappear down the far side of the ghyll. Ten minutes after the dogs were sent for them, the sheep rise up out of the ghyll close to our feet. They are beaten and they know it. They trot obediently across the moorland and join the flow of sheep heading home. The dog sees that we have them now and turns back down to its master deep below. Joe gives us a distant wave and heads off. A dog like that is worth its weight in gold. My mouth was slightly open in awe when I saw how distant it was on the skyline. I had to shut it not to seem silly. My dogs, for all their merits, couldn't have done that. We aren't easily impressed but there is a kind of respectful hush at what we have just seen.

An old shepherd turns to me and says, 'That is a proper fell dog.'

'Yes,' I acknowledge, 'but don't tell him. His head will swell.'

At the far end of the fell, I wait as I've been told. I'm not sure whether it is seconds, minutes or hours that pass there, because there is no sense of time.

I watch small trickles of sheep heading home, pushed by the men left behind me. Joe has almost cleared the ghyll out and I join up with him to cut across the far end of the fell. We pause to admire a Herdwick tup (ram) lamb that is passing us, chased by the dogs.

'Look at that.'

'Yes.'

'It is one of yours.'

'I know.'

'The mother just passed without it a minute ago.'

'It will win shows that one.'

'Maybe.'

'Time will tell.'

He cuts behind me and pushes the sheep across the heather. And I head around the skyline, pushing sheep down to Joe and clearing out the peat hags. I am the furthest point from home now. I see my world stretched beneath us, the three kinds of farmland that make up our world: meadows (or 'in-bye'), intake and fells. The farming year here revolves around the managed movement of the sheep between these three kinds of land.

A fell farm is at heart a simple thing. It is a way of farming that has evolved to take advantage of the summer growth of grass in the mountains to produce things that farmers can consume themselves, in a subsistence model, or sell to earn their keep.

Nothing makes sense without reference to what went before, and what comes afterwards. It is literally a chicken and egg thing (or a sheep and lamb thing, if you prefer). But it might help if I explain briefly the basic structure of our working year. At its simplest it works like this...

Midsummer we keep the lambs healthy, gather the ewes and lambs down from the fells or intakes for clipping and make the hay for winter.

Autumn sees us bring the sheep down from the fells or higher ground again, for the autumn sales and shows, taking the lambs from their mothers (who can then recover from their efforts) and preparing and selling the surplus lambs and ewes in the 'harvest of the fells'. In these few short weeks we make most of our annual income, from selling surplus breeding females to the lowlands and a handful of breeding males that are good enough to be sold to other breeders at a premium.

Late autumn is about starting the breeding cycle by putting the tups with the ewes, including the newly bought tups from other flocks. It is also when the retained lambs (those required for the future of the flock) are sent away for winter to lowland farms.

Through late autumn and winter we also fatten and sell our spare male (wether) lambs to butchers for meat. Our farming is largely about producing breeding sheep for sale to other farmers (who value the daughters of the fell flocks because they are tough and productive on lower ground) and male lambs for meat from the abundance of grass in the mountains between May and October. There is an intermediate trade in these lambs called selling them 'store', which has a middleman buy them and fatten them. What money we make is from these two kinds of production.

Winter is about looking after the core breeding flock through the worst weather of the year, feeding them when needed. Our sheep eat grass for much of the year until it disappears in the winter months, when we need to feed them the hay.

Late winter/early spring we tend to the pregnant ewes and prepare for lambing time.

Spring revolves around lambing the ewes on the best land we have (the in-bye), and looking after hundreds of young lambs.

Late spring/early summer we are marking, vaccinating and worming the ewes and lambs and pushing them to the fells and intakes to take advantage of the summer growth of grass, freeing the valley bottoms to grow the hay for winter.

And then we do it all again, just as our forefathers did before us. It is a farming pattern fundamentally unchanged from many centuries ago. It has changed in scale (as farms have amalgamated to survive, so there are fewer of us), but not in its basic content. You could bring a Viking man to stand on our fell with me and he would understand what we were doing and the basic pattern of our farming year. The timing of each task varies depending on the different valleys and farms. Things are driven by the seasons and necessity, not by our will.

Sometimes you are left alone somewhere on the mountain, waiting for the others, alone in the silence. Skylarks rise, ascending in song. Sometimes there are moments when not a sheep or a man can be seen. Away in the distance you can see the main roads and the villages. No one really knows how long this fell gathering has been happening, but quite possibly for as many as five thousand years.

Beneath my feet and all around me is the rough mountainous grazing land. Traditionally, Lake District farms like ours had common grazing rights for a set number of sheep on the common belonging to their manor. The numbers were fixed by custom and communal management to reflect the grazing capacity of the fell and the winter grazing capacity of the farms down below.

This was, and is, a system that requires rules and customs to prevent abuse, cheating or mismanagement. Before mobile phones or email, the only way that people could work collectively to manage this land was to have agreed traditions and practices – making it clear what everyone was supposed to do, when and how. There were even manorial courts to punish wrongdoing with fines, a practice that still exists through the commoners' associations. We collect stray sheep from each other at our shepherds' meet in November, or we are fined by the other commoners. Travelling by road from one side of a common to the other to collect a stray can be a ninety-mile, or more, round trip. Some farms still have stocks on different commons, so some fell shepherds spend a lot of their lives gathering different fells. Some farm lads specialize in this kind of gathering as an extra way to earn their living, and have packs of sheepdogs for the work.

There is a poetic fantasy that shepherds, and farmers, live a kind of isolated existence alone with nature. Wordsworth encouraged that idea, offering the world an image from his childhood of the shepherd alone in the fells with his dogs, at one with nature. At times this is physically true to life – men like my grandfather were sometimes alone with their sheep and the natural world. But it is equally true that shepherds don't exist alone, culturally or economically. My grandfather had a field called the 'Football Pitch'. There were enough young men working on the neighbouring farms that they could muster two teams for a match. And his work was about dealing with, and ultimately impressing and earning the respect of, other people.

Apparently the Bedouin can navigate the Sahara because they have an extensive knowledge of the dunes and sandy ridges, and even though these move slowly over time, they can count the ridges and know with a degree of accuracy where they are and how to get to where they are going. Our cultural navigation, our placing of ourselves and other people, works on a similar structural basis – if you understand the bones of it, you can navigate the detail.

My grandfather and father could go just about anywhere in northern England and they'd usually know who farmed the land and often who had been there previously, or who farmed next door. The whole landscape here is a complex web of relationships between farms, flocks and families. My old man can hardly spell common words, but has an encyclopaedic knowledge of landscape. I think it makes a mockery of conventional ideas about who is and isn't 'intelligent'. Some of the smartest people I have ever known are semi-literate.

If my grandfather could find out where someone farmed, the breeds of livestock they kept, and which auction mart they frequented, he could quickly find common ground with any farmer in the north of England, or even in the rest of the UK. He knew what everyone was likely to be doing at any given time of year. 'Don't bother going to see the Wilsons... They'll be too busy dressing mule hoggs [the beautiful ewe lambs they sold each autumn for breeding on lowland farms] today,' he'd say. And if you went to the farm over the hill that he was talking about, you'd see that he would be right.

Long before anyone could have a credit ratings check, people here could quickly find out if someone new in the community was trustworthy or not: a few questions at an auction mart or at a show with someone from the person's previous community, and their whole pedigree and track record would be passed on.

So someone being accused of sheep-stealing is a matter of scandal, a dirty rumour that flows through the valleys. Recently, a well-respected Pennine farming family was accused of stealing sheep from many of their neighbours. The case has not been to court yet, and I have no way of judging whether it will end in a conviction or an acquittal – but the shockwaves it sent through the hill-farming community were profound. An old man we know who farms the same common had tears in his eyes when he told us about it, as if he couldn't believe someone he trusted might be guilty of cutting out ear tags and sawing off horns with the burnt-on flock marks, and then stealing the sheep.

There is an unwritten code of honour between shepherds. I remember my grandfather telling me about his friend buying some sheep privately from another farmer for what he thought was a fair price. Weeks later he attended some sheep sales and realized that he had got the sheep very cheap indeed, too cheap, about £5 each less than their market value. He felt that this was unfair to the seller because he'd trusted him. He didn't want to be greedy, or perhaps as importantly, to be seen to be greedy. So he sent the farmer a cheque for the difference and apologized. But the farmer who'd sold them then politely refused to cash it, on the grounds that the original deal was an honourable one. They'd shaken hands on it. Stalemate.

The only way out was to go back the next year and buy his sheep and pay over the odds to make up for it, so he did. Neither of these men cared remotely about 'maximizing profit' in the short term in the way a modern business person in a city would. They both valued their good name and their reputation for integrity far more highly than making a quick buck. If you said you would do a thing, you had better do it.

My grandfather and father go out of their way to do good deeds for their neighbours, because good will counts for a lot. If anyone buys a sheep from us and has the slightest complaint about it, we take it back and repay them, or replace it with another. And most people do the same.

Fathers' names are interchangeable with those of their sons, and surnames with the names of the farms. The name of your farm tells other farmers here as much about you as your surname. There might be twenty farmers with the same surname, so it is immediately followed by the name of the farm for clarification. Sometimes the name of the farm even replaces the surname in general discourse.

I met a man in a pub recently and he knew my grandfather. 'You'll be a fair man if you are half the man he was,' he said sternly, then bought me a drink, the accrued interest on some unspoken good turn my grandfather had done for him decades earlier. Anyone new to the community or common would be watched carefully until they had shown themselves to have integrity and to play by the rules. They say you have to be here for three generations before you are a 'local' (people laugh when they say that, but it carries a high degree of truth).

Floss and Tan are working hard. They cut backwards and forwards, driving sheep across the land. Sometimes one of them will bolt off into a dip or a hollow and return with some ewes that have been out of sight. We sweep the scattered ewes and lambs across the peat hags and the expanses of heather towards Wolf Crag. I see the dogs of the man I am expected to join with. I can't see him, but they are working to his commands from somewhere unseen, so we have effectively joined up. He will have seen my dogs over the brow of the fell and will know I am here. He cuts beneath the crags and meets up with the old shepherd who is in charge of proceedings. I see them a few hundred feet below, swapping notes about how we are doing. Occasionally an arm will extend to point out some information or other. Their dogs are scattered across a wide area, working sheep homewards. The crags beneath me are steep and dangerous. If I took five steps forwards, I could easily tumble to my death. I can see for maybe twenty miles.

The first time I gathered these crags I was with an old shepherdess who I was negotiating with to take over her flock of fell sheep. We had been friends for many years, but I was being observed to see whether I could manage sheep with a dog on a fell, a kind of unspoken test. Half a dozen ewes and lambs were sticking a hundred yards below us on a grassy ledge halfway down the rock face. I sent our old dog Mac down the cliff face, through a little grassy descent between two rocks. He threaded his way down and brought them out gently but firmly at the bottom. He made me look good. She said he had done 'all right', which from her is the highest praise.

When we have cleaned out the crags, we have the sheep in one swirling mass, a woolly carpet laid over the lower slopes of the fell. The noose of men and dogs is tightening now and many hundreds of ewes and lambs are threading home in front of us. Sometimes in bad weather we lose a man, and wait patiently for him to reappear out of the clouds or mist. So we pause sometimes and wait, holding the line. Then, when everyone is done, we drive the massed flock of maybe four hundred sheep down into the sheepfolds that are littered along the lower slopes of the fells. Usually these consist of little more than a drystone wall surrounding a gathering pen, and a couple of fenced or wooden-railed collecting pens for sorting them in.

We chase the ewes through the narrow sorting 'race' (a walled alley, down which the sheep flow, with a gate that sorts them left or right into different pens at the end), where they are divided up into their own flocks. You need a great eye and fast hands to 'shed' (to swing the gate back and forth at the end of the alley), because, if you're lucky, you get three seconds to identify the sheep's flock mark and open the gate the right way. I work them into the race and shout out any badly marked sheep or lambs (occasionally a 'white' lamb will appear, that is, an unmarked lamb, having been born at the fell, and we will find its mother and thus its owner). One of the other shepherd's dogs nips my hand as I push the sheep through. I yelp and threaten to kick the dog. His owner shouts to ask what the hell I am doing.

'Your dog bloody nipped me.'
'Serves you right, yours bit me the other day.'

We both laugh. That's quits.

Soon the flocks are clear and held by their shepherds and dogs from mixing again. The mountains behind us should be empty and silent.

When the last ones are sorted, the shepherds walk their sheep home for clipping.

Everywhere is noise.

Men shout.

Whistle.

Holler.

Clap and wave hands.

Ewes call for their lambs.

Lambs call back.

Dogs bark.

The men drive the sheep away home. They fleet away like the shadows of clouds blown across the lower slopes of the mountains.

The past and the present live alongside each other in our working lives, overlapping and intertwining, until it is sometimes hard to know where one ends and the other starts. Each annual task is also a memory of the many times we have done it before and the people we did it with. As long as the work goes on, the men and women that once did it with us live on as well, part of what we are doing, part of our stories and memories, part of how and why we do those things.

In June and July, on a suitably dry day, when we were not otherwise engaged in the hay meadows, my grandfather would gather the sheep into the sheep pens. I remember this day from thirty years ago as if it were yesterday. The men are sorting the flock through a shedding gate at the end of an alley, so the lambs are in one pen and the ewes in wool in another. They then run these ewes into a building where my father clips them. My mother is trampling the wool down to pack it into the bags.

Dad's T-shirt is wet with sweat. He straightens his back occasionally as if it aches. He catches the sheep from a pen, and turns it with a twist of the neck over his leg, on to its bum. A hand reaches up and pulls the shining rope that starts the motor. The other hand tucks the ewe's leg behind his bum and picks up the clipper hand-piece. The stomach is clipped first, with a hand reaching down to protect the teats or the penis. Then wool on the back leg is opened out round to the tail and the backbone. The machine sweeps the fleece, with successive blows of the arm, from the sheep's bodies. Dad is like a machine, the sheep sort of entranced by his movement, an all-consuming dance between him and the sheep. It is a carefully choreographed thing in which the sheep is turned, shuffled and rolled in clever purposeful ways so that each sweep of the shears takes a full comb of wool from the body, with that part of the body stretched safely so there are no vulnerable clefts of skin caught and cut by the shears. The ewes are ready for clipping, so the fleece rises from the skin, lifting away from the body with the comb of the electronic shears gathering it in and the cutter cutting it cleanly and neatly from the body. The ewe loses its fleece without stress, and is back with its lambs before it knows what is happening.

Dad could shear maybe two hundred sheep in the day. He wears moccasins sown out of a hessian wool bag and rough-stitched across the top of his feet. These help him to feel the sheep and to caress them around his legs to get the cutting comb full of wool without cutting loose folds of skin. You can clip in boots, but you lose the feel of the sheep and the flexibility needed to bend in all the right places.

His motor hangs from a ladder that is jammed between two rafters in the barn. From it hangs a drive-shaft that powers his hand-piece, which is silvery smooth from heavy use. Once or twice each summer, a ewe will struggle and be nicked by the clipping machine. My grandfather would sew the wound up if it were deep, with the thick needle he used to sew up the wool bags, or, if just a nip, he'd send me up the hay mew to gather cobwebs which he would then press on, helping the blood to coagulate and scab.

A few years later, when I was in my mid-teens, I learned how to clip from my father. It felt impossible. I was awkward and clumsy and the sheep felt as if it was fighting me. I had no stamina, and my feet were not moving when they should have. My knee-bending, steps and rolling somehow not quite in sync, I couldn't find the rhythm I needed. I tried to fight through it and it just got worse.

He was always faster and fitter than me.

I felt like giving up, walking away.

It is cruel work for men.

I got tired and the sheep felt it and fought the process.

But tough work knocks the silliness out of

you when you grow up in places like ours. It teaches you to get tougher, or get lost. Them that are all talk are soon found out, left sitting, feeling sorry for themselves, exhausted by mid-afternoon, whilst the older men are grafting away like they have only just started.

Dad would look across, mid-sheep, and ask if I was tired, a taunting question. I'd feel like punching him. I couldn't keep up with him for years. I hated that, and fought it, and I got beat even worse. Later I stopped trying to race him. I found I was beating him sometimes. He got older. I'm not the fastest clipper around, but I'm not bad, I make a tidy job. After a few days to get my fitness up, I'm reasonably fast.

The ewes are plagued with flies by clipping time, and flick their ears to shake them off. Our farm has lots of trees and woodland so we get lots of blowflies and bluebottles. By July the flies are at their worst, and we cannot wait to get the sheep clipped and dipped (soaked to the skin in a chemical 'sheep dip' to repel flies) so they can look after themselves better. A handful of ewes each summer have fly 'strike', infestations of maggots. Creeping, hungry, vicious little bastards, they take hold in a soiled patch of wool, then in the flesh, or in a foot. We first know when a ewe holds up a leg, as if in pain, or twitches, or is trying to bite her side, or simply gives up on the way home and lies down. A 'struck' foot is sometimes a mass of wriggling maggots; a tail or patch on the wool is harder to spot and can spread across the body. Left untreated, they can kill and clean a sheep to the bones in a month. The flies swarm around an affected sheep, the smell making them desperate. Clipping these ones is unpleasant because flies bite your arms. A horse fly, a clegg, leaves a red swollen welt on my father's arm and he curses like a demon. My grandfather takes a 'struck' ewe to the side and pours on Battles Maggot Oil. The maggots wriggle out, abandoning ship at the smell of the noxious stuff. The floor becomes flecked with dead and dying maggots. Away to the side are the ewes in wool, waiting to be clipped. The shed is a cacophony of sheep noise as they shout to the lambs that are waiting noisily for them to emerge into the sunshine. The clipped ewes find their lambs by their calls, but the lambs often seem confused by the skinny bald creature that greets them, and rush off again to find a mother that looks right.

A good clipper can shear as many as four hundred sheep a day (some more) but two hundred is a respectable score, and would break most people. My father would some-times help neighbours in a gang. A team of four men can shear well over a thousand in a day. This requires a whole bunch of other people to gather the sheep, sort the lambs off, push the ewes on to the clipping trailer, wrap the wool, mark the sheep after they are sheared, lead the batches of sheep away, and generally keep things moving. It's the time of year when tempers are short, the buildings alive with the hum of the shearing machines, sheep baaing, dogs barking and men shouting. Some years are a bloody nightmare for shearers, as sheep should not be clipped wet, so you have to try and get them into barns before it rains. But many are gathered into specially erected pens in the fields, and sheared on mobile clipping trailers, so the weather can ruin a day. Today we use electric clipping machines, but it is still bloody hard graft, and as many helpers as possible is a good idea. Lots of young, and not so young, shepherds earn their keep through the summer in gangs of shearers that travel from farm to farm doing the work. Farmers' wives still compete with each other to put on the best clipping time tea (no one has the heart to tell them that being full of cakes and scones is not great when you have to bend double all afternoon).

The only thing wrong with clipping time is that wool, one of the great products of the world, is sold for so little. Once, wool was a key cash crop from farms like ours, a major part of the income. They say caravans of horses or donkeys led bales of wool across the fells to Kendal (which was built on the wool trade) until late in the nineteenth century. Much of the wealth of the monasteries that owned much of the Lake District in the Middle Ages was created from wool. Today, if we paid someone to shear the sheep, it would cost about £1 per animal. The fleece is only worth maybe 40p, just a token payment against the costs, not a profit.

Some years we don't bother to sell it because the price is so bad, and just burn it. Herdwick wool is wiry, dark and hard (which makes it ideal for sheep on mountains and for tweed jackets, insulation, or carpets that last for a very long time, but less than ideal for competing with other man-made products). Look at old pictures of Herdwicks and you will see they had more wool than they do now, because farmers respond to market incentives and have bred sheep with less and less wool: we clip to help their welfare, not earn a living. But my grandfather would still scold me if I didn't tug off the 'dags' of dirt, or failed to pick up the 'lockings' (handfuls of loose wool) from the floor.

As my father releases a clipped ewe, he throws the fleece to the side. My grandfather sweeps it up and casts it like a fisherman's net across the wrapping table. The fleece lies for a second like a coat inside out. He pulls any dirt from it and picks out any straw or twigs. He rolls the outsides of the fleece inwards so the fleece forms a foot-wide rug. He rolls it into a ball, starting at the tail end until he reaches the neck. Then, with a pull and a twist, he turns the neck wool into a kind of rope. In one movement he would bind the rope around the bundle, and tuck it firmly beneath itself on the other side of the roll. The fleece now was tied and would be thrown to my mother to be stuffed firmly into the corner of the wool bag. When I was small and too young to work, I'd be in that wool bag, greasy with lanolin. I'd just lie there as the shed hummed to the sound of the motor on the clipping machine and the sheep. I can remember lying there, looking up at the swallows coming and going to the nest on the beam above me as if nothing was happening. The young birds occasionally peered over the edge to watch the commotion. Sometimes I'd fall asleep in this woolly cocoon, to be woken later by my fussing grandmother who then plied me with short-bread or something else she had baked. She'd spit on a hanky and rub my face clean. My grandfather marked the ewes with our farm's 'smit' mark as they were released. Ours is a blue mark in front of red on the sheep's shoulders: it tells everyone they are our sheep.

A few days later, we would dip our sheep. The ewes start to resist at the slightest smell of the stuff. So we would have to manhandle them into the dipping tub. They'd be tossed into a grey chemical soup that repels flies, then swim around, looking for a way out. One of the men would dunk them with a long staff with a metal prodder on the end. Us children would go down to the river and admire the dead fish downstream from where the trickle of dipping flowed, their upturned twisting bellies flashing silver in the stream. No one worried about such things too much back then – but basically we were dipping them in chemical agents developed to kill people in the First World War.

These are long hard days. They start early with bringing the sheep into the yard. The sheepdogs toil hard to gather them up. I remember my grandfather working his dog like this on clipping days. He was struggling to move fast enough, but he had a great sheepdog. His dog Ben was a beautiful, strong-boned black and white Border collie, a strong dog that could work a big flock of sheep. He even trained Ben to catch a single ewe on command without hurting it, holding the fleece, without nipping the skin, and using his strength to anchor the ewe until Granddad could hobble closer and grab it. But Ben was cheeky, he knew he couldn't be caught by the old man, so he would taunt my grandfather by bouncing in front of him as they went to do some work, and my grandfather would shout blue murder at him.

F this. F that. Threatening the evilest of punishments if he caught him.

Ben just bounced and smiled.

But once the work started, Ben would focus and together they could do almost anything. After he'd worked well, all the cheek was forgotten and never mentioned again, until the next day when they would repeat their act. Later, when my grandfather was older and had a stroke, we made a bed for him in our front room at the farmhouse. We brought Ben in to see him, and he was so happy to see his beloved sheepdog that he cried.

A black lamb breaks back past me and bolts off up the road. I shout to Tan to go and fetch it back. He heads off after it with his long loping stride and passes it in a few seconds. At the point where he reaches it, dog and lamb are side by side. Tan kind of nudges it off-balance with his nose as they gallop and it tumbles in the grass and turns over. It comes back to the flock, parting the foxgloves and thistles by the roadside. I breathe out, because a lamb can, if it panics and decides its mother has been left behind, go all the way back to the fell with its head down, oblivious to dogs or men.

I had eaten my sandwiches at the fell gate, and the day had cooled. Clouds appeared in the western sky. Goldfinches trilled excitedly as they flitted from one patch of thistle fluff to the next. The long straight road falls away in front of me. Then the lanes take me down through the 'allotments' or 'intakes'. This is privately owned or farmed land on the lower slopes of the fells or on the moorland (common land that once was divided up, so the commoners each had an 'allotment' of land). They are often rocky, heather-covered, semi-scrubland and steep. The intakes look similar to the fells but are divided by snaking dry-stone walls that reach up the fell. Many of these fields were enclosed from the seventeenth century onwards, and were often used for grazing cattle. Unlike the common, these rough fields are farmed by just one farmer.

Taking my sheep down those lanes is what people have done here since the land was first settled. That is what these lanes, or 'outgangs', are for, to let the little farmsteads access the mountain grazing. I am walking in the footsteps of my ancestors, and living a life they lived.

The farm I am heading to back down these lanes was, and in some ways still is, my grandfather's farm, bought in the 1960s. It is also my father's farm: he kept it going, paid for it and added to it with extra land in the 1970s and 1990s. It is also my farm, because I've worked on it with them both since I was a child, and because I have built on it a new farmhouse and buildings and taken my family there to live. I will spend the rest of my life keeping it going.

The farm we are returning to with the flock is already partly my three children's farm too; they share in its day-to-day life now. They have their own sheep in the flock, so they can start to build them up and learn about the highs and lows of farming. They are expected to work with me as I did with my grandfather and father.

Their sheep are called Moss, Holly and Loopy Loo. Who am I to argue? It was the same for me when I had two sheep called Betty and Lettuce. It goes on.

Some people's lives are entirely their own creation. Mine isn't.

The sheep I am walking back, bought after my test from my neighbour, make it a true fell farm with its own fell-going flock. The sheep she had taken on in the early 1970s (from another noted breeder) were handed on to me. The flocks remain; the people change over time. Someday I will pass them on to someone else.

As my grandfather did, you can farm here on your own land in the valley bottom without taking sheep to the common land on the fells, by farming 'improved' sheep breeds that don't need to be as tough. He farmed Swaledale ewes and bred hybrid North Country Mule lambs to sell each autumn at the big sales at Lazonby in the Eden Valley. The farm he bought had no fell grazing rights for sale with it. He wasn't really a 'fell shepherd'. He was one step down the mountainside from the fell flocks, buying lambs from the fell farms or selling them tups. He thought that just fine, because further down the hill was better land and better sheep if you were a progressive mid-twentieth-century farmer as he was.

Swaledales are tough moorland sheep with thick wind-turning fleeces and bold black and white markings on the face and legs. These are, originally, as the name suggests, the sheep of the Pennines, but have become almost universal in the uplands of northern England because they have the ability to breed an incredible hybrid daughter (if mated to the improbable-looking Blue-Faced Leicester) called a North Country Mule, a wonderful sheep with speckled brown, or black, and white faces, and perfect petticoat fleeces, that go down to the lowlands and provide the breeding flock for the rest of the UK. Swaledales are widely farmed in the Lake District. My grandfather kept these to produce lambs, which he sold each September. And because he produced these cross-bred lambs, he had to buy in new 'draft' ewes to refresh the Swaledale flock each year.

These daughters of the mountains are the best commercial ewes money can buy for a lowland farm. They inherit the hardiness and maternal instincts of their mountain mothers, but also the 'improved' growth rate, body and fine fleece of their lowland fathers. After their youth in the mountains, they also do extremely well on almost any other type of land across the UK, because everywhere else is an improvement. They are a rich productive harvest from these farmed mountains. So farmers descend on these little auction marts in droves, until the lanes leading to them are choked with traffic. The stereophonic din from the auctioneer echoes out across the pens and over the surrounding fields. The air is full of the smell we love, the scent of the dipping that crimps their fleeces and colours their wool the brown tea-stained shade that tradition dictates. Their black and white speckled faces are scrubbed sparkling clean, and little bits of red and blue woollen thread hang from their necks to show they are the selected 'top pen' or the 'seconds'.

Farmers from elsewhere have bought the surplus breeding stock produced here for many centuries, as the northern fells are a kind of nursery for the national sheep flock. My grandfather sold sheep each autumn to farms as far afield as Somerset or Kent. It has long been a trading economy: a thousand years ago we were part of a Viking trading world that stretched north around the Atlantic coasts.

Each autumn, farmers from lower, kinder ground buy the spare ewe lambs from the mountains for their flocks, and the male lambs to fatten for meat. This movement is a simple necessity because the mountains can carry many more sheep in the summer than in the winter when the grass disappears. The mountains produce a vast harvest of breeding sheep, meat and wool. In addition to the lambs sold, because they are surplus to the fell flocks' needs, thousands of young sheep from the mountain flocks were, and still are, wintered on lowland farms, with their owners paying by the week for their keep. These sheep go back to the fells the following spring to become the future of the flock, in time for the mountains to turn from blue and brown into summer green.

But in the past decade or so, my father and I have deliberately made our farming system more traditional and old-fashioned, returning to a system with minimal external inputs and expenditure, because it helps us escape from the spiralling costs that are killing small farms like ours. And because we have slowly learnt that the traditional ways still work.

Taking those steps has been an education, taking us into the commons farming life of the fells, and it has led us to learn a great deal about the system that has survived there. Our land is not one inch closer to the fells than it was thirty years ago, but our relationship with them has changed. I am still learning about this landscape.

For the last half a mile before I reach home, I am following lanes flanked by drystone walls, lined with pink foxgloves and ferns. I am now passing between the fields of my neighbours. There is no common land down here. Lake District farms like ours tend to have a small amount of privately owned or managed 'in-bye' land or pasture in the valley bottoms, divided by drystone walls, fences or thorn dykes, giving it that patchwork 'green-and-pleasant-land' effect. This is our land, owned by us, or rented, and where we have to grow any crops we need for winter and where we can best look after young lambs on the ewes in spring. These meadows are vital to the working of a fell farm, and were created so that winter could be survived here.

A huge amount of work was invested in making this place farmable, a lot of which was undertaken in the twelfth and thirteenth centuries: clearing the fields of trees and boulders; taming the becks and channelling them so they drained the land and didn't wash away the topsoil with each flood; and building the walls and boundaries, grabbing new bits over time from the forest and scrub, draining the boggy valley bottoms. Without the walls, hedges and fences, this land would have been grazed all the time with no hay made for winter. It would have been a boom in summer, then a bust in winter. Lack of fodder would have brought starvation for cattle and sheep and, ultimately, people.

As I pass down the lane, I see a wall that I helped build with my grandfather.

I remember him teaching me to wall, starting me off aged maybe eight years old, filling in the gaps in the middle with poor little stones, whilst his mole-like hands faced the wall with hard blue walling stones. Summer is a time for repairs and maintenance – making good on the damage of the winter passed.

The American poet and sometime farmer Robert Frost wrote a fine poem about mending walls:

Something there is that doesn't love a wall,
That sends the frozen-ground-swell under it,
And spills the upper boulders in the sun;
And makes gaps even two can pass abreast.

It is true here that 'Good fences make good neighbours'. My grandfather knew that and wanted me to know it too. I watched him turn the stones over in his hands, looking for walling edge, and place them each in turn into the gap: the plain and unloved side of each stone to the interior of the wall, and the 'walling face' to the outside edge. He placed some 'through stones' across the wall to hold it from bellying out in the years to come. He'd encourage me to backfill behind them with smaller stones, using my smaller hands to make it solid and wedged in place with fist-sized lumps of slate and rock.

He patiently saved some of the best stones for the 'coins' on the wall top, and placed them back as they had been before with the silver, yellow and sun-bleached green mosses and lichens facing the sky once more.

Once, some people stopped a car to take photographs and he turned and walked away. He murmured 'bugger off' under his breath. He regarded the tourists that swarmed past on sunny days as minor irritants, like ants – they got in the way and they had strange ideas, but a little bit of bad weather and they'd be gone again to leave us to get on with stuff that mattered. He found 'leisure' a strange, modern and troubling concept – the idea that anyone would climb a fell for its own sake was considered little more than lunacy. So he suffered tourists, but found them incomprehensible. I don't think he understood that those people had another perception of ownership of the Lake District. He would have found that as odd as him walking into a suburban garden in London and claiming it was 'sort of' his because he liked the flowers.

Much of the day-to-day work on a farm is spent on the hundreds of little un-newsworthy jobs that are required in managing the land and sheep. Mending walls. Chopping logs. Treating lame sheep. Worming lambs. Moving sheep between fields. Running sheep through the footbath. Laying hedges (only in months with an 'R' in them, or the sap will not run and the hedge will die). Hanging gates. Cleaning the rainwater gutters on the buildings. Dipping sheep. Trimming sheep feet. Rescuing lambs from being stuck in fences. Mucking out the dogs. Trimming the muck from the tails of ewes and lambs. As you drive past, you wouldn't notice them, but they add up over time. Landscapes like ours are the sum total and culmination of a million little unseen jobs.

The sheep ahead of me have stopped now, having met some walkers coming the other way. The visitors thread through the sheep looking nervous and pass by me. They say hello. So do I. Then they pass onwards, one of them clutching a copy of a Wainwright guidebook.

I wonder whether any of them see the wall my grandfather built, or care that it stands, or wonder who built it.

We are nearly home now.

The sheep can sense it. Some of the older ewes have threaded ahead. They fan out to graze where the lane opens up by a stream. They are reluctant to cross it and pause on its banks. I send Floss past them with a curt command of 'Away'. She nudges through the lambs and away past the ewes and jumps the beck. I tell Tan to 'Lie down' and he holds the back door shut. I walk past the flock and open the wooden gate on to our land. It is held with a rusting length of barbed wire twisted tight. I untie it and swing open the gate. The eldest ewes know they are back to our farm, their other home, and start to jump the beck and thread into the field. Five minutes and they are all back on our own land. They find their lambs and head off to graze.

Floss and Tan lie down in the beck, wallowing with only their heads out of the water, their long pink tongues panting. Blue-green dragonflies zip to and fro above them.

My grandfather had nearly been crippled as a boy with cerebral palsy osteoporosis. The doctors had said he would never walk again. For months he was pushed around in a wooden wheelchair. But after several weeks in a care home in Carlisle and some wonder drugs, he had slowly recovered. I remember watching him dress and seeing a hole through one of his white legs where it had eaten lumps out of him. They say that after months of being spoiled and doted on by his mother Alice, he was 'wasted something rotten', 'spoilt' and 'impossible'. He'd got a sense of himself from his mother, and illness had freed him from being broken into an obedient son. He was, after that time, never really likely to play a secondary role to his father. Like lots of lads in places like this, he would have to find his own farm. So when he was about twenty years old, he borrowed money from his mother to take a rented farm in the Eden Valley from the Lowther Estate.

The Eden Valley is a wide fertile plain that stretches on its eastern boundary from the Pennines, the backbone of England, across to the Lake District fells in the west. It stretches from the Solway plain in the north, and the city of Carlisle, down to the Howgill Fells and the Yorkshire Dales in the south. It is, and has long been, famous for being a place of great sheep and cattle, with some of the best grazing land in Britain. If you stand in the sandstone villages in the fertile plain at the bottom of that valley, you might think you are simply in a lowland area, a place for arable farming or dairy cattle, a place disconnected from the mountains in the distance. But it isn't – it is bound to the mountains through the movement of the sheep downwards each autumn. The fells and the wide lush river valley are all part of one ancient interconnected farming system.

The farm my grandfather rented was nestled below one of the limestone ridges that fall away from the eastern Lake District fells. It was exposed, windy land. To work there could leave your face chapped red like it had been sandpapered. From the highest fields, 900 feet above sea level, you looked over the valley, stretching miles away from you below. It meant we were kind of halfway between the fells and the valley bottom. The old men said the farm was no good, too steep and hilly, it 'would wear out good horses'. Maybe he was foolish and lucky, or maybe he was wise, but soon horses disappeared, replaced by tractors, and this age-old constraint no longer mattered.

Tough farms were not places to get rich, but they offered opportunities to those willing (or forced by necessity) to take a chance: the young, the keen, the poor, the proud, maybe the foolish. If you had a big lowland dairy farm with good soil, you probably looked down your nose a bit at these farmers on marginal land. These tough farms are two months behind in the growing season and don't even clear their meadows of sheep until into May, by which point the lowland farms just ten miles away are almost ready to mow their grass. The timing of everything, from lambing time to making hay, is determined by where you farm and the quality of your land.

My grandfather worked hard and turned it around. He supplemented his farm income by working on other neighbouring farms. He was a good horseman. He dealt in livestock and was an opportunist, like so many of his peers. If pigs paid, breed or fatten pigs. If Christmas turkeys paid, fatten turkeys. If selling eggs paid, get hens. If wool was wanted, grow wool. If milk paid, milk cows. If fattening bullocks paid, buy bullocks. Adjust, adapt, change. Do whatever you needed to – because you stood on your own two feet, there was no one to pick you up if you fell down. The geographic constraints of the farm are permanent, but within them we are always looking for an angle.

His pride and joy was a beautiful horse that pulled his trap called Black Legged Boxer. You could see it shining black in the sunshine, muscles rippling, as he described it years later. But there were bad times too. He suffered a disaster when his horses died of grass sickness. He still sounded devastated forty years later when he told me about it. He would drive into the village of Mardale to buy sheep (now deep under the dark blue water of Haweswater reservoir, a valley drowned to supply Greater Manchester with water), or to sales at Ambleside or Troutbeck in the Lake District to buy sheep from the fells.

To buy sheep profitably, you need to be smart, aware of the price of them in different markets. He'd turn up in the farmyards and would be invited to inspect their lambs, feeling their woolly backs to see how much meat covered their backbones and ribs, and judging whether they would grow well. Then he'd make them an offer. He had to be able to see the opportunity for profit over these hill farmers who didn't often leave their valley. But he had to try and be fair as well, because otherwise you couldn't come back. He had to know how to manage livestock to get them ready for the markets when they were most valuable, and how to buy stock that would 'improve' on his land: taking sheep on to worse land would ruin them, and his profit.

Soon there was a family to fill the white-painted farmhouse, and by the 1960s he had worked and grown the farm business to the point where he could borrow money to buy his own farm. He borrowed £14,000 and bought a run-down and badly fenced hill farm in a little valley called Matterdale. The farm he bought, which is now my home, was a step backwards in some ways, even from the quality of his rented farm. Hilly fields, brown patches of sieves (rushes), thistles, small fields, and surrounded by fells that seemed to anchor rain clouds. Hardly an efficient farm for an age that demanded that farming must become ever bigger and more 'efficient', but it was what he could afford.

This land required a different kind of farming, with different breeds of sheep, because it had a shorter growing season and more rain. But he knew that owning a small hill farm was more secure than renting one. Owning his own acres gave him freedom and a secure asset that could grow in value. You could be 'put off' a rented farm by your landlord. So my grandfather took us back into the hills, taking his chance. He kept the rented farm and farmed it as a unit with his new owned land in the hills. There was a farmhouse for sale as well, but he couldn't afford that, so he farmed the land from a distance at first, and then later built a bungalow next to the old barns and sheep pens.

It is quite normal to have bits of land scattered many miles apart. The land directly next to the land you have rarely comes up for sale. So it is not at all unusual that my grandfather bought a farm fifteen miles away from his rented farm.

By the 1980s, my grandfather had made it a fairly smart farm, with good livestock. He was proud to be a good 'stockman' devoted to his stock; he could buy and sell things wisely, and was a good judge of breeding. At a glance, he could see little things wrong with sheep or cattle, like if they were wormy or short of minerals. Some faults meant they would lose money; others could be easily cured and transformed into profit. Men like that can judge the weight of a lamb by sight and roughly cost in a few seconds how much it will take to fatten a bullock. He knew just when sheep were going 'stale' and needed a change of pasture.

Being smart is held at a premium amongst farmers. You very publicly live by your decisions, as judged and measured by those around you.

I sometimes think we are so independently minded because we have seen just enough of the wider world to know we like our own old ways and independence best. My grandfather went as far afield as Paris for a trip to an agricultural fair once. He knew what cities had to offer, but also had a sense that they would leave you uprooted, anonymous and pushed about by the world you lived in, rather than having some freedom and control. The potential wealth on offer counted for little or nothing set against the sense of belonging and purpose that existed at home.

As on a Scottish croft, there often isn't a full wage available for a farmer's sons and daughters starting out, so many do something else for a few years to earn their keep until there is a wage for them to live from the family farm. Many of the older farmers, too, often worked for decades off their farms – in the mines, on the roads, splitting slate, walling, clipping sheep or just working for someone else. It is still common for young farmers to do all sorts of other work to make ends meet until they become the 'farmer' – and the parish records confirm this was always the case. Farms were just too small to keep everyone.

Unfortunately, it was normal for farmers like my grandfather to borrow money. It is how they bought land they could scarcely afford ('Well, they won't be making any more of it so we better had'), and this meant they had a connection to the bank and the rest of the world through interest rates. So the waves of prosperity or hardship that affected farming here were often created by global events like the world wars, the Industrial Revolution, the Great Depression or the massive expansion of farming in the American West in the nineteenth century. It is a sad logic, but wars were generally considered by the old men to be 'good for farming'. Events like the Napoleonic Wars or the world wars had stopped the cheap imports that undermined our way of life, and reminded politicians that producing food at home mattered. Then afterwards they'd forget and things would get gradually worse. But in later times, too, we were often sucked into the wars that affected the rest of the world.

There is a little cemetery on the Somme full of local lads, many of them farmers' sons, who enlisted and died together in July 1916. One of my grandmother's uncles had been shell-shocked, recovered enough to come home and work on the farm, but had broken down a few years later when working in the fields with his brothers. He had lain down, sobbing in the brown soil, surrounded by turnip leaves. They had carried him back to the farm, and later he was taken away to a lunatic asylum in Lancaster. The war went on for him for his whole long life. He was still spoken of affectionately when I was a child, because visiting him was a fresh memory.

Some of my friends' families still farm on the Lowther Estate in the Eden Valley that their great-grandfathers had secured from 'Lordy' because they had fought for his Lonsdales (a battalion in the Border Regiment) in the First World War. Before the war, Lord Lonsdale had been a friend of Kaiser Wilhelm, and when the war broke out the tabloids teased him about his loyalty, so he went into recruitment overdrive to show what a patriot he was. The War Office had to stop him recruiting 'undersize' farm lads from the local valleys. My grandfather's uncle was one of the best shots in the regiment. My son now carries his name, Isaac.

As soon as I could toddle, I'd be packed into Granddad's Land-Rover and we would go off to do a job on the farm. My mother would be left fretting whether I'd be properly looked after, and about what he would feed me. Once he rushed back and told my mother I 'really really needed a jimmy-riddle' (his word for a pee). He couldn't get me out of my dungarees. For some reason I never figured out, my granddad always called us, 'Us two old men'.

> 'Us two old men will go and get them sheep.'

I'd ride with him looking out of the open Land-Rover window. One day the door wasn't latched properly and, when he braked, I swung out, holding on to the window for grim death until I was rescued from mid-air. I vaguely recall escaping other painful and imminent deaths at auction marts, clambering up the railings to escape wild cattle or feeling the air whoosh past as a bullock kicked out and just missed me.

The world I was born into extended between our two farms; the outer limits of civilization were friends who did similar things to us, some as far afield as the Pennines or the valleys of the Lake District. Beyond that, the rest of the world didn't really exist for most of my childhood.

I was always curious about other places, but I had no desire to go to any of them. And we didn't do holidays. Instead, I'd be packed off with my granddad to stay on their farm, where I would follow him around all day, or climb into their bed at night.

Early summer has quieter moments when we all breathe a little easier after the slog of lambing. My grandmother and grandfather would 'take the sheepdogs for a walk'. This was a walk for maybe a mile up the lane beneath the fell where they lived. In reality it was all about enjoying the sight of their farm in its summer mode of abundance. The ewes and lambs grazed contentedly below. The fells shimmered in the dusk, red, orange and blue. The hay fields below like patches of purple with their flowered grasses headed up. You could almost taste the sweet hay smell, the pollen in the air. The valley echoed always to the sound of ewes calling their lambs.

There is a rusting metal gate halfway up that lane where they would stop and lean on the gate and look over their home. The setting sun would bathe the valley, the fields a golden haze of insects, thistles and flowering grasses. I'd listen to them talk, and feel their love of, and pride in, this place.

Some summers we would walk our cattle up into an isolated little valley called

Dowthwaite Head, to graze land belonging to a farmer my grandfather knew called Mayson Weir. Once we got there, the cattle would graze away across the fields, flies and dust rising around their feet, their tails swishing in the sunshine. They'd be 'fat as butter' when we went back for them in the autumn. The movement upwards of sheep and cattle enabled the best land below to be protected from grazing to make hay.

Mayson was a 'character'. You could get lost up there in his whitewashed farmhouse, having a whisky. 'Go on, you'll have another.' Then it was too late to argue. My grandfather's glass would be three fingers deep with whisky. I'd sit listening to them joking and swapping stories and gossip. I'd munch custard creams or gingerbread. I remember them telling of a shepherd who had died when a bunch of them had gone to a pool to cool down on a hot summer day. He had dived in and not come back up. Sometimes Mayson would disappear and come back with some home-cured bacon from his back kitchen. It would be covered in the fuzz that grows on it. He'd cut some off and produce a 'fry up'.

Today, thirty years later, I am friends with Mayson. Families like ours roll on beside each other, through the ages, with the bonds enduring. Individuals live and die, but the farms, the flocks and the old families go on.

When my grandfather bought our farm in the fells, he took us into the landscape of another breed of sheep, the Herdwick. Herdwicks are born black with white ear tips, but change colour as they age, until they have a white, hoar-frosted head and legs, and a blue-grey fleece. They are arguably the toughest mountain sheep in Britain. Snow. Rain. Hail. Sleet. Wind. Weeks of dour wet weather – no problem. At one day old, with a good mother, they are almost indestructible, regardless of the weather, with a thick leathery skin and a carpet-like black fleece that keeps them dry and warm. The ewes can live on less than any other sheep in these conditions and come off the fells with a lamb of value in the autumn. Recent scientific research has shown that Herdwicks are genetically rather special, they have in them a primitive genome that few other British sheep carry. Their nearest relatives are in Sweden, Finland, Iceland and the northern islands of Orkney. It is believed that the Herdwicks' ancestors lived on the islands of the Wadden Sea, near the Frisian Islands, or further north in Scandinavia. Local myth has it that they came with the Vikings on their boats, and the science now suggests this is true. Since they arrived they have been selectively bred for more than a thousand years to suit this landscape.

The first time I saw Herdwicks on our farm was as a child. Somehow they had more character than more modern sheep. The six-month-old lambs stood there watching me knowingly. Dark-brown fleeced, with sturdy white legs, a touch of the teddy bear about them in their early winter coats. My grandfather had bought a hundred from a neighbouring farmer to fatten them. He was rather startled by them, as (unlike our other breeds) they seemed to think his modest farm was paradise and got fat rapidly and he was able to sell them for a quick profit. They had probably lived their whole lives until that point on some of the rockiest and most unforgiving mountain terrain in Britain. Like a lot of farmers in the twentieth century, we were keen to keep the most modern 'improved' sheep we could on our land, and our modernizing farming worked OK in a world of abundant and cheap fuel, fertilizer, feedstuffs and cheap labour. But keeping improved breeds on land that is a bit too tough for them is hard work: they ail more, eat more, grow slower than they should, and ultimately die more often. Later, as oil and feed prices rose, and the cost of the inputs on our farm increased, we would learn that native breeds like the Herdwick are the best suited to a life with little supplementary feed in this landscape. But at that time, Herdwicks seemed to us noble, though like a thing from the past.

Herdwick sheep are sometimes mistaken for a 'rare breed', or believed to continue to exist purely for nostalgic reasons, or because of Beatrix Potter and the National Trust (on account of the Trust's role in buying some of the farms here to conserve them and through that conserving the fell flocks in situ). Herdwicks aren't rare: with more than 50,000 breeding females, they are still the only commercial sheep for the highest Lake District fells. They are currently experiencing a renaissance of interest as farmers try to find ways to farm tough land with fewer inputs, and because people are waking up to the quality of their traditional, naturally produced and tastier meat.

Each breed of sheep has its own community of breeders that come together at different auction markets for the sales. So my grandfather knew and dealt with farmers from across the Swaledale country: from the Lake District in the west to Durham in the east, and from the southern Pennines to the Scottish borders. The auction marts are in many ways the centre of our way of life, the places where we come together to trade, but also to socialize. When I was a child, they were situated where they always had been, in the centres of the local towns, with sheep and cattle still walked in from local farms. In the last thirty years, they have mostly moved to the outskirts of the towns on to industrial estates – in the name of scale and modernization. But I think something important has been lost through this, a link between people who live in towns and our world.

The different breeds have their own calendars, with everything from lambing to clipping being carefully timed to fit the annual growing cycle and to ensure that they are in peak condition for the autumn sales, which are mainly breed-specific.

My grandfather went to these sales and bought lambs from the high Lakeland fells in the autumn to fatten during the winter on his better land. These lambs are called 'stores'. He sold them a few weeks later, when they had put on weight and condition, for a profit at a 'fat lamb' sale. I can remember being taken along when my grandfather bought them, when I was about knee-high to the grown-ups, to the little auction mart at Troutbeck. It is just over the brow of the fell from our farm, a mile or so as the crows fly. It was little more than a tiny, wooden, octagon-shaped shed, topped with a corrugated tin roof, surrounding a ring where the sheep were driven for sale, surrounded in turn by acres of pens holding thousands of sheep. The ring was a sea of sawdust, surrounded by wooden seats where the buyers could congregate, facing the wooden rostrum of the auctioneer. The sheep stood outside in long rows of pens made of wooden or metal hurdles (gates). Drovers, some still wearing clogs, and all with sticks or flapping plastic feed bags, would bring the sheep to the ring, or walk them away. As each pen was bought in the ring, the buyer's name would echo down the pens from one drover to the next until it was scratched in chalk on the little blackboard on the pen gate. On wet days, the heat of the sheep would result in a damp, wet-woolly smell. Steam would rise from their backs.

They'd push us kids to the front through the legs of the old farmers, or palm us off on the older kids to look after, and bung us some Fruit Pastilles or a Mars Bar to keep us quiet. We'd sit munching a gob full of chocolate, watching thousands of sheep being sold whilst our fathers and grandfathers did their business. I loved listening to the old men talking. One of them was my grandfather's cousin, and they said he had been to Oxford University when he was a youth. I remember thinking that was a strange thing for an old farmer to have done.

Making hay. Clipping. Looking after the ewes and lambs. Gathering. This is what summer means to us.

Making good hay is like a commandment from God if you live here. People would once have faced ruin or even famine if they couldn't feed their animals through the winter. Misjudging your crop, even now, is an expensive gamble that can wipe away the year's profit in an instant. They say that about once every ten years it was virtually impossible to make hay. It would rain and rain and never let up. So my deciding to be born at hay time meant there was more than one thing of importance happening.

I arrived on a sultry July day as my father and grandfather were making hay in the meadows on our farm, trying to beat the impending rain that every farmer fears as he secures the crop needed for winter. Grass that dries in the sunshine, and is then baled and stored in the barns, makes for wonderful winter fodder. It bursts out of its bales on a snowy day when the sheep need it and you get a breath of summer. Even the flowers in the meadows can be seen pressed in there. But hay that is rained upon starts to rot. A little rain makes hay that is a bit like grassy strands of cardboard. The ewes will eat it in winter and it will keep them alive, but it is not the same. A lot of rain and the hay becomes a rotten, inedible and bitter-smelling mess. Eventually it is useless.

My mother had struggled to get pregnant, until she received some – then innovative –

fertility treatment. Though her grandmother always insisted that she'd actually been cured by riding a horse of my grandfather's and having her insides 'jiggled about a bit'. Our council house was half a mile down the road from the village, in a row of four. It faced on to our fields, was grey-rendered and fronted the road. In the photos they look surprisingly fashionable in a 1974 kind of way, Dad with his wavy hair, sideburns, wide-collared shirts and tight trousers with flares; and Mum pretty and always looking as if she worshipped me. Mum and Dad look, in those same photos, like extras in a *Jaws* movie. Longer hair. Dreamy looks.

The house was wallpapered with hideous 1970s patterns. They didn't have much. But in the pictures they look very young, and really happy. Dad has a bit of mischief in his eyes. They say Mum was always reading me books. Mum says she can remember being hustled out of the way with her new baby (me) when it was time to feed the men on the farm. Grandma lived and breathed being a farmer's wife. Good meals on the table when she said they would be. Nothing should get in the way of that. Later, Mum had a fridge magnet that said 'Dull women have immaculate homes'.

The day before I was born, my cousin had come to stay so Mum could babysit him, but during the night she felt something starting and walked the mile to the phone box in the village (the council house where they lived didn't have a telephone). The rather stroppy matron she got on the phone told her to stop panicking, I wasn't due for six more weeks, go back to bed and stop being such a silly first-time mother. Fine. She walked home and went to bed. The next morning my dad disappeared to work, mowing the hay. My mother took my cousin home in the car. When she got there, my auntie was concerned about her and took her to our local little hospital. She was rushed to Carlisle hospital in an ambulance. My auntie called in the offices of her husband, a solicitor in the local town, to get him to go and get my father from the hay fields. He rushed out in his car and his pinstriped suit, waved my dad down on the tractor, handed him the keys to his car and told him to get a move on to Carlisle Hospital. I was coming.

My uncle was left standing in his polished shoes in a dusty hay field miles from anywhere, not sure what to do. So he drove the tractor fifteen miles back to the farm. Half an hour later, my father screeched into the car park at the hospital and came to find Mum. I should have been born in the autumn. But I was healthy and strong. My first day intertwined with the happenings on the farm. The first time I saw my dad he would have been in his work clothes – dusty, sweaty and smelling of summer hay. Once I was born, he went back to get on with making the hay (when the elder of my two younger sisters was about to be born, he took my mother to hospital via a field that needed shepherding and nearly didn't get her there on time).

Some of my earliest memories are of summers in the hay fields following my grandfather around. I would be sitting or sleeping behind the forever-bouncing seat of a tractor whilst someone else baled or turned the hay. Once I was mobile, it was a time for running and leaping over rows of hay, building dens in the bales, fishing in the streams that cut through our meadows. As long as the sun shone, it always felt like a special time of the year – as if all was well with the world because the cattle and sheep could look after themselves, for the most part, for a few weeks in summer, and we were going to have the crop to feed them over the next winter.

Hay times were like chapter markers in my life, each one showing me to be a little stronger and more useful and my grandfather a little older and weaker. I literally grew into his shoes. In the good summers, or perhaps just in my memory, there was an air of joy about it, and my grandmother would come to the field at regular intervals with meals or afternoon tea, cakes she'd baked and a large tin jug of tea. We'd sit around on makeshift chairs made of bales, and the old men would tell stories and joke about summers past. I loved those stories about working horses, the heroic labours of men in the past, and the German and Italian POWs who had come to work on the farm during the Second World War.

Granddad didn't reckon much of the Italian officers, their claims of aristocratic pedigrees, or their somewhat different work ethic. They were all Count-bloody-this ... and Count-bloody-that.' And they wolf-whistled at girls passing by in the train carriages. Some of those POWs were still living on the farms they'd chosen to stay at after the war, rather than return to a home that didn't exist anymore. They lived in little bedrooms in farmhouses all over our area, like strange living ghosts of the war that had ended before my father was even born.

The wind would catch wisps of hay in little tornadoes and whizz it off across the field. Swallows would hawk around the field catching insects. High atop bale-laden trailers, I rode home, dodging branches and telephone wires. Once, the trailer caught a gate post as we turned into the yard. I tumbled down on an avalanche of bales, landing at my grandmother's feet. She clucked and fussed. The men denied, perhaps truthfully, any knowledge of me being up there. I just shrugged.

The hay meadows were criss-crossed with shadowy little streams, flanked with foxgloves, havens from the heat of the day for sheepdogs and children. These meadows were not mown until late summer, so that the flowers and plants could drop their seeds. Traditional upland hay meadows are a thing of beauty. Rich multi-coloured waves of grasses dancing in the light summer winds. Mosaics of brown, green and purple grasses and flowers are home to a multitude of insects, birds and occasional roe deer calves. Lush, green, thistle-scattered pastures flank the hay fields, with the twin-rearing ewes watching the commotion with interest. Grasshoppers call to each other from the ribbons of green that are the field boundaries, and magpies chatter from the crab apple trees.

In an ideal world, hay timing would be easy. Three or four days of perfect drying weather after the grass is mown, two or three sessions of turning it to ensure it is uniformly dried by the wind and sun. The hay, dry and sweet-smelling, would be baled and then led into the barn, never so much as touched by a drop of rain. But it is not like that very often in English summers. Timing the mowing to hit the gaps between rains is a calculated gamble at best and a bad summer can ruin a winter's fodder, and often did in the Lake District. So hay time is often a battle between the farmer and the weather.

Cutting the meadows leaves the mower covered with a thick carpet of grass seeds, pollen and insects. It also opens up a hidden world where voles had lived in peace, but now scurry off to the dykes. In one of our meadows, the sun-bleached skeletons of two elm trees stood, from where a kestrel would watch us work, occasionally hovering above the field and swooping down on a vole and carrying it off in a fistful of talons.

Following the mower, perhaps a day later, would come the haybob, which fans the grass out of the rows in which it lies and helps it to wilt evenly in the sun and wind. For the next few days, sand martins sweep past us as we turn the hay each day, scattering insects to the breeze.

When the greenness and sap have wilted out of the grass after a few days, it is rowed up ready for the baler. And at last the baler starts thumping out its dusty clunking rhythm. The men work under the keen eyes of greedy lice-tormented rooks, which wander the fields searching for worms and grubs under the cleared rows. From time to time a 'sheer-bolt' might snap on the baler and you would hear frenzied hammering and a few 'fucking hells'.

Today, hay time is increasingly mechanized (in the 1980s new machinery came in which meant that crop can be wrapped in plastic even in damp summers and some nutritional value is saved by pickling it as 'silage'), but throughout my childhood and youth it was a full-on physical effort with everyone expected to pitch in.

Once the bales were made, they had to be taken to the barns and eventually man-handled into the 'mews'. Stacking bales was one of the jobs we dreamt of being strong enough to do when we were boys. Each slow year of growing up was filled with the hope that, next year maybe, we would stack bales with the men. Our family had a shortage of young men, so we looked enviously across the fence to our neighbours who could muster a full gang. As each bale was hand-lifted several times before it was put in the barn – and we made thousands – strength mattered.

Each year I found myself a little stronger, and able to lift the bales higher, while my grandfather grew weaker. His sense of his own decline was only eased by the pride he had in me, his grandson, growing up to take his place. As a child, I had rolled bales to his knees, thinking I was helping, and carried his bottle of cold tea from heap to heap, wishing I was as strong as him. And each year the balance between us altered in my favour. Then we reached a curious halfway point where we worked as equals, when I was about thirteen years old, but I quickly agreed each time he suggested that 'us two old men' ought to stop 'for our pipe' (neither of us smoked). The next year I was much stronger than him and pretended I needed to stop, every now and then, so he could rest. A couple of years after that, he was following me around the field, rolling bales to my knees for me to lift, and lifting the odd one when he could.

Making hay in daydreams tends to be idyllic and sunny, but in real life it can be a bitch of a thing. I can remember 1986, the worst summer, when we burnt all our hay. A disaster. You need nearly a week of dry and sunny weather to make hay. And you need to be able to travel on the meadows with a tractor and mower to mow the grass at the start of that week. What could possibly go wrong in one of the wettest places in England?

In 1986 it just never stopped raining. Black clouds. Miry fields. Endless rain. Sometimes summer never quite happens. It must have offered brief moments of respite, though, because somehow we got the hay baled, but then the heavens opened and it rained for days and days. If you understand the importance of good hay, there is something irretrievably sad, pitiful and pathetic about ruined hay. What should be a lovely sun-bleached green slowly becomes grey, rotten and dead. What should have been our harvest for the winter rotting into something worse than useless, a time-consuming liability. We tried stacking the bales against each other, on days when the wind blew and rain eased. But the bales now sagged dead-weight beneath the baler twine in your stinging hands. More rain. Fat splashing drops. The heaviest I've ever seen. The hay was ruined. It had started to sprout green on the tops of the bales. It would never dry out. Everyone knew it. Even if we got it into the barns, it would 'heat'. It might even combust and burn the barn down as sometimes happened on farms. Or it might simply rot. There was no point in bringing it in. Rooks skulked in the ash trees, waiting for worms under the heaps.

The fields were now green with 'fog' (the sweet regrowth after cropping that we use for the lambs that are weaned off their mothers in August and September), the bales sulking and leaving rotten dead marks where the grass should now have been cleared. However bad it was, the hay needed to go somewhere. Clearing the fields of this sodden junk was like moving corpses. Cruel work. Sickening for men. Pointless. Rotten smelling. We took thousands of bales to the ruins of an old stone barn, created a fire-beneath one corner of the pile. Stood back and watched. But the cursed stuff couldn't even burn properly. It smouldered sulkily for weeks. I can still smell the hay burning in a stupid, pointless, charred heap. We brought bales to the heap for days until the fields were cleared, sweating, with rain dribbling down our necks. When we were finished, we had nothing to show for weeks of work or a year's growth on the meadows. No hay in the barns. Fields now boot-deep in grass except for coffin-shaped dead yellow stains where the bales had lain. My father turned away and said, 'Never mention this to me ever again, I don't want to remember it.' Grey smudgy clouds anchored to the fells and it rained on for weeks.

I followed Granddad around for years. Like all good grandparents, he could only see the best in me, and that always made me stand a little taller. I was the 'squire' in training, so he taught me the things he thought I needed to know to farm here. Practical little things like how to build a wall, or how to prepare a sheep for sale, or how to judge stock, but he also taught me values and how to think about things, how to deal with people fairly, and earn respect, how to do business and how to protect our good name. It was instilled in us from the start that we were part of a family and a community and had values to uphold, and that these were more important than our own whims and fancies.

The farm and the family came first.

We are, I guess, all of us, built out of stories. He told stories of his grandfather on his mother's side of the family, T. G. Holiday. From what I could gather, my granddad had worshipped and copied his grandfather much as I did mine. So even though I'd never met this man and he died long before I was born, there is a connection and continuity between us. My grandfather built himself up out of stories about T. G. Holiday and I built myself up out of stories of him.

T. G. Holiday sits proudly on my bookshelf in a sepia photograph I have inherited. It must date from the 1890s or 1900s. The picture shows him standing in a field surrounded by bullocks, a hazel stick in hand, and large droving dog sitting loyally by his feet. Bowler hat. Mutton-chop whiskers. He looks deep in thought and not that interested in having his photo taken. The cattle are eating from wooden buckets or stone troughs. In my grandfather's stories he was a kind of mythical heroic character.

T. G. was a tenant farmer on the Inglewood Estate. He bought Irish cattle and met them in the little harbour at Silloth with his men. He'd have wagons loaded with troughs for feeding them on the journey back to the farm. He also bought flocks of geese off those boats, and tarred and gritted their feet so they could be walked home. They walked these animals back, taking a couple of days or more, sleeping at nights by the roadsides. He fattened the cattle and geese on his pastures and sold them in local markets when they were in peak condition and worth the most. Without anyone much noticing, he made money from his dealing, because during the First World War he quietly accumulated lots of War Bonds as an investment. Some time later he sold them for profit and filled two suitcases with money, which he kept at home for two years.

Then, one day, he loaded his suitcases into his cart and went to an auction of three good farms. To the disbelief of everyone present, he bought all three farms and paid in cash. As he travelled home that day, he passed a crowd on the roadside in Penrith and realized they were selling a row of cottages with sitting tenants. Perhaps to finish off whatever point he was making that day, he bought the row of cottages with the cash left in the suitcases. He sold them to the individual tenants for a profit in the next few months.

If you were a tenant farmer and wanted to make your mark on a small farming community, then you couldn't do much better than T. G. Holiday did on that day. He was thereafter someone of standing. He set each of his sons up on the bought farms, educated his daughters (including my great-grandmother Alice), and helped them get started with their own husbands. He is still known and spoken about with respect by older farming folk. His descendants in several families are still proud of him, generations later. A lot of farming families have stories like this, their own myths of how they came to be who they are.

My grandfather had an eye for things that were 'beautiful' like a sunset, but he would explain it in mostly functional terms, not abstract aesthetic ones. He seemed to love the landscape around him with a passion, but his relationship with it was more like a long tough marriage than a fleeting holiday love affair. His work bound him to the land, regardless of weather or the seasons. When he observed something like a spring sunset, it carried the full meaning of someone who had earned the right to comment, having suffered six months of wind, snow and rain to get to that point. He clearly thought such things beautiful, but that beauty was full of real functional implications – namely the end of winter or better weather to come.

From the beginning, my grandfather taught me the classic worldview of what Europeans would call a 'peasant', and we would simply call a 'farmer'. We owned the earth. We'd been here forever. And we always would be. We would get battered from time to time, but we would endure and win. There was also a strong sense of what others would call 'egalitarianism' that exists in many pastoral communities in Northern Europe that judged a man or woman on their work, their livestock and their participation. Historically, there had not been the wealth to differentiate between farmer and farm worker in these valleys, at least not in ways that divided them socially and culturally. The aristocratic families didn't, or couldn't, really exert their power here, and there was little idea of 'class'. The men, farmer and labourers, worked together for the most part, ate at the same table, drank together in the pub, watched the same sports and generally lived very similar lives. The farmers who owned land perhaps thought they were a little smarter than those who had never managed to get a farm of their own, or the farm workers, but any form of snobbery or class distinction was fairly alien. You couldn't get away with being a snob. The world was too small. There were too many chances for others to make you pay heavily for it. Respect was mostly linked to the quality of a man or woman's sheep or cattle or the upkeep of their farm, or their skill in their work and management of the land. Men or women who were good shepherds were held in the highest esteem, regardless of being to modern eyes 'just employees'. To be a shepherd was to stand as tall as any man.

I went to a really good little primary school. But my bookish mother and school didn't stand a chance. I knew from the start that school was just a diversion from other things that mattered more.

But it wasn't all wasted. I had a magical teacher called Mrs Craig who read me *I Am David* (about a little Jewish boy escaping a concentration camp). She also read us the *Odyssey* and I remember loving the bit about Odysseus and his men clutching to the bellies of his giant fat sheep to escape the One-Eyed-Giant's cave. I still love these books. The teachers said kind things to my mother about me being 'bright' and 'enigmatic'. But the bottom line was I belonged to the farm.

My grandmother once scolded me for idleness when she caught me reading in her house. The gist of it was that there couldn't possibly be so little else of value to do on the farm that I could justify reading a book in daylight hours. Books were considered a sign of idleness at best and dangerous at worst. My school successes (increasingly rare, as I got older) also seemed to worry my grandfather, like a flashing warning light that he might lose his heir to another culture. There was nothing much useful in books. School had to be attended. But it was just a dull obligation.

I remember a school night in the hay field, in an eight-acre banked field called 'Merricks'. It was five minutes past the dictated curfew for me. But I was a little man, too busy for homework and books and all that stuff. Nine years old. Working with them that counted, with an itchy neck, stinging hands and prick-led legs. Then on the skyline, a car caught the red sunset, a familiar Ford Sierra trailing dust down the lane. 'Quick!' said one of the men, and he pointed to the half-stacked heap. 'Get in the middle.' I jumped into the heart of the stoop between two bales; half a dozen other bales were tucked around me. As I was entombed, I watched the car reach the field gate through a peephole left open between bales. The old man chuckled as he laid on the 'top 'uns'. I could hear the car roll up the grassy stubble... 'Have you seen him?' All I could see was the fly-specked bonnet of the car, my heart pumping, in my grassy tomb. 'Nope.' Then silence for a telling grown-up moment. 'Well, it's past his bedtime, and it's a school night.' 'Reckon I'll tell him, if I see him.' 'Reckon you should.' The car crept away back home, and I watched it through my peephole.

My grandfather worked, but he also played hard and drank hard. Tuesdays were auction days. It was an all-day thing with a bunch of other established farmers (farm workers and sons did the work at home). After the sale was done, they'd end up in a pub, pissed. Word would get around the womenfolk and eventually the men would be hunted down. A disgruntled wife or two would arrive and drag her man out of the pub. Once, I picked up one of their shepherd's crooks from the pub floor where a drunk had knocked it down and he gave me £5 for 'being a gentleman'. My grandfather seemed to know everyone and be on good terms with most of them. He'd been up to mischief with them all at one time or another.

He passed on stories inherited from his own grandfather that spanned back and forth across vast periods of time as if the 1850s or 1910 were yesterday. The 'silver' and 'brass' my grandmother polished included things brought home by soldiers in the family from the Boer and Crimean Wars.

Granddad could read and write, and everyone in our world thought he was smart, but there was only one book in his house and it was about horse ailments. It's fairly safe to assume he had never read Wordsworth or Melvyn Bragg. What use were books and schools to this man?

My grandfather was aware of the modern world, and could adapt to it. But he also held its values and new-fangled inventions at arm's length. He would return from the auction mart and ask my mother, who was 'educated' (one term at university in Norwich before meeting my father and chucking it in), to work out the figures 'on the computer'. The 'computer', which he didn't entirely trust, was a small battery-operated hand-held Sony calculator. Intellectually we were, in short, little more than 'peasants', with a classic, small 'c' conservative world-view inherited through an oral tradition, based on stories and passed-on wisdom and experiences, and yet we existed within 1980s Britain as everything changed around us. If you removed the tractors and machinery from the farm, much of what we did, and how we did it, was ancient.

Granddad even called things by ancient names, like 'Mowdies' for moles, or 'Mel' for the post hammer we used and 'Gaeblic' for the iron pole you make post holes with. He called for the 'yows' (ewes) with strange shouts that didn't mean anything to modern ears.

'Hoeeew Up, Hoeeew Up.'
'Cus, Cus, Cus, Cus.'

Years later I watched a TV documentary about reindeer herders in Sweden and one of them called the reindeers in a very similar way.

He had a wicked sense of humour and mischief was always bubbling beneath the surface. There was a touch of the Brer Rabbit about him. He could handle himself. I can remember officials from 'the Ministry' (of Agriculture) coming to talk to him about the 'biodiversity' in our hay meadows and what they expected him to do to manage those meadows for the flowers and birds in return for the subsidy they paid. After an hour and a half of observing him nodding and agreeing to everything they suggested, they departed, and I asked him what they wanted. He said, 'No idea... The secret with them daft buggers is to say yes to everything they want, and when they've gone carry on regardless.'

My grandmother was a farmer's wife of the old-fashioned kind. Women like her were once everywhere in our countryside, working behind the scenes to feed armies of men and playing an important role in the farming of the landscape. In the Lakeland valleys the women often did the farming while the men went to earn wages in the mines or elsewhere. My grandmother kept her house and the garden spotless, weeding with an old mother-of-pearl butter knife. No weed was safe from her for several hundred yards around her farmhouse.

I once drove into a farmyard with my dad and he noticed the wild yellow poppies growing out of the wall, 'Your grandmother wouldn't have liked that... She would have thought this place was being let go.'

Grandma was 'long-suffering', as the cliché goes. My grandfather was a 'character', and all agreed that at times he would have been a proper bastard to live with. There was some stable girl he'd got pregnant decades earlier. This was common knowledge, but never spoken of. It was swept under the carpet where it remained as a visible lump.

So theirs was not like Hollywood-movie love, more like the love between crocodiles. His idea of fun was chasing her across the kitchen, trying to grab her around the waist and cuddle her, while she swiped at him with the frying pan and called him a 'dirty old man'. He'd wink at me like it was a lesson in woman-handling.

Maybe I had the blindness of a grandchild, but it felt like she enjoyed the fighting. I felt it, despite the words. Sometimes it looked like hatred, sometimes just like love.

They had been through a lot together and had had a 'good life', albeit full of troubles and incidents. When I worshipped him, he was getting old, and frustrated that old age was beating him. But he was still full of mischief. Theirs had been a shotgun marriage, and not the only one in our family judging by the dates of first-born children in our family tree. Grandma's stories often featured lost babies or children that had died of TB or polio or in accidents on the farms.

Grandma polished the 'brass' with Brasso like our lives depended upon it. She fed pet lambs from old Schweppes lemonade bottles with worn red teats, and collected dog food in an old pan on the kitchen worktop, then soaked it in milk. The stale smell of cold beef and potatoes was the smell of the back kitchen. My abiding memory is of her bent double, apron tied tight around the middle like a roll of muslin tight-bound with string, angrily chiselling out weeds from between the cobbles, or in the kitchen turning out meals for the whole family.

Bacon and egg, sweet from the dripping left in the pan too long, egg yolks freckled with old fat. Toast carpeted with butter and syrup, cut at a funny angle. Rice pudding like a cake, rich and creamy, with a caramelized brown halo around the dish's rim. Fish and chips wrapped in old newspapers, delivered to the door of our dens in the hay or the woodshed. She baked every week: rock buns, apple pie, shortbread. She considered it a grave insult to her housekeeping skills if a visitor wouldn't have a cup of tea and cake.

Their home was like my own home, somewhere I was doted on and spoiled. My earliest memory is being about five years old and lying in bed with them, because I wouldn't sleep in my own bed, playing with and comparing their ears. On the wall hung a tapestry of Jesus that I hated, but could never work out why. 'We Love Him Because He Loved Us First' it said. There was a little ornament on the sideboard of an old woman sewing that looked just like my grandmother, and an owl with a broken porcelain ear. We smashed that, by mistake, and she nearly cried.

She didn't really understand new things like TV. She barely even tried. She lived in a world that kind of died sometime in the 1970s and 1980s, a world that felt as if it stretched from the beginning of time until then, when a woman was judged on her cooking, her house and her garden. She didn't understand the emerging new world of the 1980s, our world, of books, money, computers, credit cards and holidays. She lived with an unquestioned belief that these things were aberrations, foolishness and fleeting fancies, the rubbish of now. So she taught us good rules that no longer made sense. Rather than understand our brave new world, she closed her eyes tight shut and turned away. When my life changed in my twenties and I had to briefly become something else, the shared understanding between us broke. We were like foreigners to each other. I hated that and missed her.

When I was in my late teens, my mother and aunties organized a 'pizza party'. It was quite an event – our first foray into 'foreign food' (a takeaway ordered from the newly opened Italian restaurant in our local town – and, yes, this is only twenty years ago). Grandma was horrified. She arrived looking quite cross, like we had all forgotten ourselves. She was convinced that 'pisa' was a terrible new-fangled idea, and that we might all be poisoned if we ate such 'rubbish'. She declined to taste a slice, her face scrunched up in disdain, and thought we had all gone mad when we tucked in and enjoyed it. However, she recovered some pride and struck a blow for England when she pulled out a tin of freshly baked shortbread that she had smuggled into the party. When we had eaten that, she went home convinced that her shortbread had seen off the challenge of foreign food forever.

'I don't know anything,' she'd say when asked about the past. The secret was to work her into the subject, until she'd talk and talk. She had taken in evacuees in 1940, and still turned up her nose at their manners fifty years later. Some tramp had arrived on the doorstep, all high heels, lipstick, fur coat and no knickers, but this urbane creature had soon evacuated back to the bombing zones, deciding it was safer to face the Luftwaffe each night than my grandmother's scowls. But her tone changed when she told sad stories of a young POW from Hamburg, who had worked on the farm and shared their table, her cataract-clouded old eyes twinkling some story that I couldn't read.

When she was old, she had a kind of altar: a table of photos, with all of us, her tribe, in shiny silver frames, in christening dresses, wedding dresses and racing colours. The mantelpieces were littered with silver tankards and cigar boxes with inscriptions of long-forgotten racehorses that my grandfather had trained on the farm, with magical names like Pentathlon and Cool Angel, names that raced through our lives, hooves pounding: knick-knacks and paraphernalia of the glory days, when it all made sense. A stallion carved from some dark wood pranced beneath the TV. She'd sit in the car at Cartmel horse races, sending her runners to the Tote. And on the way home she'd buy us all fish and chips with her winnings.

She would tell tales about her 'Momma' that no one really listened to or understood. Later, when my grandfather was gone, she had a little flat of her own. She poured glasses of whisky and told stories of her man. She loved to talk about him after he had died. And he glowed in those stories, like some great dead king.

One school night, I was walking with my dad across one of the meadows to check on our ewes before it rained. He stopped suddenly. Told me to 'Be quiet'. Then crawled forward for twenty yards and kind of pounced like a fox, with his cap in his hands. He smiled me across. He'd caught a leveret, a baby hare. It was nestled gently in his flat cap in the grass, one of the most beautiful things I had ever seen. It looked up at us with deep glassy eyes, and screamed. We let it go, and it sloped off out of sight. Towering, dark cotton-wool clouds were gathering all about us, and away to the Pennines thunder and flashes of lightning. We ran back to the Land-Rover with big fat raindrops soaking us.

I dreaded the thought of going to secondary school. Our little village school was full of kids like me, their fathers were friends with my father, and often their grandfathers too, all the way back. There was a scattering of non-farm kids amongst us, but they played games I had no interest in Dungeons and Dragons, and they were fashionable with new trainers and stuff. But secondary school was ten miles away in the local town. It might as well have been another universe.

I remember asking another kid on the first day what his dad did, and being told to 'Fuck off and mind your own business'. I was now in a place with different rules, where being me was a liability. Being a farm kid here was something that got you hassled, or labelled a 'yokel'. Even getting to school was a pain in the ass. The lads from the village where the bus departed used to steal your schoolbag and throw things out of the window. This escalated for weeks until I grabbed the smallest of them and punched him a few times on the floor between two seats. Because of that, some of the others decided I was 'all right'. I could pick on other people with them instead of being bullied. One day the school bus had to be stopped because a dart was thrown down the bus and cracked the windscreen. When we got to town, I was liable to get a smack from someone else because I was in the bus gang. The whole school had a brutal 'Lord of the Flies' gang thing going on.

History lessons at school didn't really go the way I hoped they would. We never did any kind of history of us, or our landscape. I think the teachers might have been surprised at the idea that people like us had a history of any interest. Instead, we studied the History of Native Americans. This was, I now realize, potentially very interesting, but then it left me confused and disappointed, and I'm not sure our history teacher actually knew anything about the subject. We also studied the Second World War and the Cold War briefly, but in such a tedious way I quickly lost interest. I remember being given a page of A4 with a cartoon showing the difference between capitalism, fascism and communism. It was hard to tell what was wrong with communism or why they might be pointing atomic bombs at our house, or why we had a hand-turned air-raid siren in our back kitchen.

When I remember the 1980s, I think of how shit that school was. It tested past breaking point all the well-meaning stuff they tell you when you are young, like 'stand up to bullies' or 'report them to the teachers'. Brilliant idea if you want a good hiding from some big lads from uptown. The lads two years above us were mean fuckers. Some were rumoured to be in the National Front and were 'known to the Police'. They lorded it over us and some of the teachers by intimidation and by ganging up on anyone dumb or brave enough to annoy them. There was no goddamn way I was going to get on the wrong side of them.

One afternoon, when they pushed through the bus queues (just because they could), everyone stepped back to let them through except some kid next to me called John. He murmured under his breath 'Fuck this' and stood his ground. The older lads looked a little shocked but still circled around. I'd never seen anyone quite as brave as this kid. He stood about six inches shorter than the older lads, his fists clenching for the fight. I wished I was him, or brave enough to help him – but my legs were already unconsciously backing away. 'I'm not scared of you,

you are a bully,' he said to the biggest lad who was striding towards him. 'I've as much right to be here as you.' I'd seen this kind of thing in movies. The underdog puts up a brief fight and the bullies back away, having learnt a valuable life lesson. For a brief second I thought that might happen. The lads paused for a second. Then the big lad pulled him over by his bag strap, so he stumbled on to the tarmac. John got a half swing away as he tumbled over, but he had no chance. The biggest lad laid a few punches into him. The other lads piled in and kicked him on the ground a few times. A few minutes later, John's blazer had a sleeve torn open at the shoulder. His mouth was trickling blood. He was doing his best to retain some of his dignity as they laughed their way down the tarmac. He still looked proud, but he now looked much younger and seemed to be shaking.

But it wasn't all grim northern clichés. Each August, my father's cousin would bring her family for their holidays to the farm, bringing a caravan or two and her parents, her husband and three boys. We didn't do holidays, ever. So I used to look forward to them coming because it made my life seem briefly like a holiday. Inevitably, they were more modern than we were. Her husband worked in a nuclear power plant as a computer expert.

And they worked all year to come and have a holiday in the Lake District. They would go fell walking, swimming, sailing, running, to pubs for meals or for picnics in beautiful settings. They were on some 'Swallows and Amazons' vibe. They crossed 'Wainwright' off the list. They would sail on the lake with their own dinghy, and later they would windsurf. They would barbecue, drink beer in the evenings and play board games. They would head off each day for an adventure in the fells, or to visit a ruin or something. They were fun and kind, and a bit different from us. They'd take me with them sometimes, as we didn't do any of this stuff. If there wasn't something happening on the farm, I enjoyed going with them, though often I was needed on the farm, so I'd stay behind. They'd often pitch in to help with the seasonal work. I was the farm cousin that showed them stuff, frogs in a wall, nests of birds I'd found, or how to do farming things like put a wall back up. Dad and Granddad were a little more detached. They didn't have much time for 'messing about'.

Maybe once or twice a summer we'd trek up a mountain, with me not quite ever having the right fell-walking gear (usually clad in T-shirt and trainers or farm boots). We'd pass people on the path wearing enough specialist climbing gear for an ascent on

Everest. I was never sure which fell we were climbing, because, out of our own valley, we didn't know their names. My southern cousins knew far more than I did about the fells because of the guidebooks.

I remember holding one of Wainwright's guides, of the Eastern Fells, in my hands as I sat with my visiting cousins on a crag somewhere up above Ullswater. The crooked lake stretched silver beneath us, glimmering in the sunshine.

If my grandfather was invisible or little more than white trash in our teacher's sermon in 1987, then the high priest of her belief system was another old man of similar vintage: Alfred Wainwright. And now I had his book in my hands. I'd never seen one of these before, because we didn't really think of the Lake District as a place written about in books, or for leisure. I'd only ever once climbed a mountain for pleasure with my parents. We went for a picnic, and a gust of wind blew away the paper Womble plates my mother had brought. Dad hadn't wanted to do the picnic anyway, they had a minor row, and we retreated back to the farm. Fell walkers we weren't.

Wainwright created a series of hand-drawn and hand-written guides for walkers that explained each of the mountains of the Lake District. Originally these were produced and self-published as a hobby, but they became cult classics in Britain and beyond, selling millions of copies. Each guide gives the reader an overview of the landscape, a series of views, some navigation advice on what can be seen from the summits, and accounts of the 'natural features', 'ascents', 'summit' and 'the view'. Thousands of people every year follow in Wainwright's footsteps up the mountains.

They are beautiful, thoughtful little books and exert a powerful hold on how other people see our landscape. They cast a spell over people like my teachers at school, whose entire perception of the Lake District was shaped by these books and a handful of others.

So I was looking down at the landscape farmed by my father's friends and cross-checking it against the guide. It struck me powerfully that there was scarcely a trace of any of the things we cared about in what Wainwright had written. Apart from the odd dot on the map for a farm or a wall, nothing from our world appeared in those pages. I wondered whether the people on that mountain saw the working side of that landscape, and whether it mattered. In my bones I felt it did matter. That seeing, understanding and respecting people in their own landscapes is crucial to their culture and ways of life being valued and sustained. What you don't see, you don't care about.

It is a curious thing to slowly discover that your landscape is loved by other people. It is even more curious, and a little unsettling, when you discover by stages that you as a native are not really part of the story and meaning they attach to that place. There are never any tourists here when it is raining sideways or snowing in winter, so it is tempting to see it as a fair-weather love. Our relationship with the landscape is about being here through it all. To me, the difference is like the distinction between what you felt for a pretty girl you knew in your youth, and the love you feel for your wife after many years of marriage. Most unsettling of all to me was the discovery that people who thought about this place in this way outnumbered us by many hundreds to one. I found that threatening to our very existence in an age when we increasingly had to do what we were told by politicians and the general public, but no one else seemed much concerned. I told my dad it was weird that none of these book people were much interested in what we did. His response was, 'Don't tell them, they'll only ruin it.'

We are on the tarmac, waiting for school to begin. We are bored, so are kicking each other, or our schoolbags. Some girl starts yelling at one of the lads. He has drunk out of the water fountain and she's yelling that he'll die if he drinks that. She says it is radioactive. We all stare at her as if she is crazy. She is one of the smart kids that will soon leave for the local grammar school and we enjoy winding them up. Chernobyl had blown up a day or two before and, according to this girl, its radioactive waste was heading our way in the clouds.

The lad at the water fountain looked a bit shocked. Then he smiled wickedly and drank some more of the water. She shouted at him, telling him he was stupid, that the clouds were spewing out radioactivity. Then we all ran around in the rain with our mouths open and our arms wide open like sycamore seeds, filling our mouths with rain. She declared us idiots, her eyes aflame with fury.

In the weeks to come, we'd learn that those clouds had deposited radiation across our mountains, and government inspectors came to the wettest fell farms to test the sheep. Restrictions were imposed on moving some sheep from the worst-affected land for years. You don't really expect men in white suits with Geiger counters in hand to turn up on your farm. It added to the general impression of my youth that the wider world was stupendously fucked up.

The whole time I was at school, I wanted to be at home on the farm. I was convinced then, and I still am, that home was a more interesting and productive place to be for me. Making anyone do something they don't want to do with thirty other bored kids seemed to me absolutely pointless. I'd look out of the windows and watch the swifts rising above the town, their scythed wings glistening in the sunshine.

One morning my grandfather caught a badger alive in a snare (he was trying to catch a mink which are an 'invasive species'). He wanted to release it. He told my dad to pick me up from school on the way to help him rescue it. Dad knew you couldn't walk into a classroom and take your kid off to do things like this. So he didn't. Granddad told me all about it later that night, how they'd released it and it was fine (but not before it had tried to take his leg off). My blood boiled. I'd been sat bored in a classroom all morning whilst someone tried to teach me Esperanto.

It felt like the whole modern world wanted to rob me of the life I wanted to lead.

The idea of wanting to escape is a strange one to me. But for the young Alfred Wainwright, it really was 'grim up North'. He had had a bellyful of grimness and wanted out. So he worked hard at school and was noticed as a bright lad with some potential, which enabled him to avoid becoming a mill worker, like his sisters, at twelve years old. He was channelled instead to more schooling and then into white-collar clerical jobs at Blackburn Town Hall. He was a classic working-class self-improver; he knew he wanted out of his situation and he worked out how to do it, and then put his head down and worked towards his goal. Later, he studied to become an accountant at the Town Hall where, amongst other jobs, he was responsible for making Poor Law payments.

Having climbed out of 'Worktown', Alfred drifted apart from the boys he had grown up with, and their mill-and-factory, rough-tongued English. He joined the ranks of the great uprooted, but educated, English middle class. He suspected, rightly, that his old friends thought him a 'snob'. But he made new friends, middle-class ones, who read books and did middle-class things like climbing, walking and daydreaming of adventures in foreign lands. But you can't help feeling that he was always a little isolated in his new world, never quite fitting in – a little lonely, his cleverness like a millstone around his neck. Some of these new friends had been to the Lake District, and read books about the Alps and Himalayas. They were fully signed up romantics, daydreaming of escaping Blackburn for the mountains.

In 1930, aged twenty-three, he made it the sixty or so miles by bus to the Lake District. Middle-class young people all over Britain were the foot soldiers of a movement that eventually gave almost everyone in Britain the disposable income and leisure time to have adventures in other people's landscapes all over the world. Wainwright loved the place he had discovered. The Lake District was all about escape for him. He never pretended otherwise. Escape from a crappy industrial and urban working-class existence, and later, escape from an appalling domestic situation (his first marriage was stunningly awful).

Later, he would engineer a move to Kendal where he managed the accounts for the town council. This enabled him to walk in his beloved Lakeland mountains in his every spare moment, and to write and illustrate his grand project, 'A Pictorial Guide to the Lakeland Fells'. It was to become one of the most curious writing and publishing adventures in modern English literature. His Guides sold millions of copies and he became a TV celebrity – the Old Man of the Fells. He 'empowered' millions of people to take to the footpaths and climb the mountains. He created a new way of experiencing the Lake District, with people now ticking off the fells he wrote about – 'doing the Wainwrights'.

All through my childhood, my auntie and uncle farmed just a mile up the road. We worked with them on seasonal tasks like making the crop. They bred good sheep, and my earliest memory is of them outdoing us. My grandfather would take it very badly, because 'all of his ducks were geese', as they say round here. But I loved to go and work with them in the autumn or go to the sales with them. I thought they did some stuff better than us, so I figured I'd learn from them and beat them later.

One Saturday in August, we were stacking hay into the mews of the barn so it was safely stored for the winter, when they appeared in the yard. My parents and my auntie and uncle left me filling the elevator with petrol and they went into the farmhouse kitchen. This struck me as strange. About ten minutes later they came out. There was something in the air, something dark and unsaid. I looked questioningly to my father and his look said 'Don't ask now'. So I didn't. We just worked.

My auntie lugged bales on to the elevator that chugged away sending bales up into the eaves of the barn, to a 'mew' of hay that grew ever upwards. She was surrounded by petrol fumes and dust. I was up in the rafters taking bales off the elevator and throwing them to my father. Light streamed through the ridged pinnacle of the corrugated roof. Sweat. Itchiness. Cobwebs. Big fat brown moths would flit about your head. The smell so sweet and dusty it could make you sneeze. My dad was strangely chatty. My auntie caught my eye a few times, and smiled. When we were done,

Dad thanked her for helping. She smiled, got in the car and went. Then they told me.

She had come to tell us that she was going to die. She didn't want anyone to see her get ill, to deteriorate. She didn't want anyone to feel sorry for her. She didn't need anyone's pity. I was forbidden to go and visit. I never saw her properly again, just a blurred glimpse of an ill woman in a car, speeding past as I worked by the road one day putting up a wall.

They say that schooldays are the best of your life, but that's bullshit. I couldn't wait to leave. I had nothing invested in it. And by the time I was fifteen, the teachers were not going to lose any sleep over unburdening themselves of me. You can't push water up-hill. You were allowed to leave after the Christmas on your sixteenth birthday, but you needed your teachers to sign you off. All any of us wanted was to get the hell out, so we envied the lucky bastards that had had their birthday and strode off across the play-ground with a white piece of paper in their fists. I have never seen most of those lads again. Today you might exchange mobile phone numbers, or stay connected through Facebook or Twitter, but these things hadn't been invented then, and few of us wanted any lasting connection anyway.

My mother had given up on my schooling by that stage, resigned to the inevitable. I more or less stopped going to school after Christmas at fifteen years old. I'd managed about a year more than my dad or my granddad. When I stayed at home, I worked, a much-needed extra man on the farm. I worked hard, so no one was bothered that the formalities of school were being ignored (I lied as well so they didn't really know what was going on). I hadn't done a thing at school from about the age of twelve anyway, just fucking around. I chose my GCSE subjects so I could be in the same groups as some girl I fancied. For the last year or two before I left, I worked part-time at home, before and after school and at weekends. A rough shout from my dad told you it was time to get up and go out and work. Feeding or mucking out cattle, or up the fields feeding sheep, and at school-bus time my mother would come looking for me, and my father would say he didn't know where I was until it was too late. One day I saw her crying as she went back to the house. Dad would give me a cheeky smirk.

I went back for some of the exams to keep my mother happy, but missed others. I showed little interest in the ones I attended, but I can remember using the quiet in the exam hall to think. I knew it was dumb to fail exams, but I preferred to fail badly by that point than have anyone think a C was the best I could do. Somehow, despite fucking about, I still got a C in Religious Studies and Woodwork, which made my granddad laugh. 'You'd make a vicar maybe...You can do the funeral service and then bray the nails into the coffin.' I'd confirmed his suspicion that school was a total waste of time.

The school had a major fundraising campaign to buy computers, and they arrived just as I left. Until then, the only computers I had seen were those in my cousin's bedroom and in the careers office at school. We'd been sent to queue outside the career adviser's door for his enlightened advice about our future professional lives. He was very proud of his careers software package and earnestly asked me a series of tick-box questions. Single-finger typing. Do you want to work inside or outside? Outside. Do you want to work with people or animals? Etc. After quarter of an hour of this, the computer started to vibrate and then spurted out a slip of paper. It said I should be a 'ZOOKEEPER'. As my dad said, when I told him, 'Bloody hell! The stupid bastards.' Then he rocked with laughter and couldn't stop.

This crappy, mean, broken-down school took five years of my life. I'd be mad, but for the fact that it taught me more about who I was than anything else I have ever done. It also made me think that modern life is rubbish for so many people. How few choices it gives them. How it lays out in front of them a future that bored most of them so much they couldn't wait to get smashed out of their heads each weekend. How little most people are believed in, and how much it asks of so many people for so little in return.

So leaving school was the best thing that ever happened to me. I felt a sense of elation that spring and summer. I was fifteen and I swore the day I left school that I would never let myself be trapped in a place like that ever again. I was going to live on my terms.

At least, that's what I reckoned.

My grandfather was seventy-two years old when he had a stroke. After a while they put him in a care home. He couldn't speak properly. It seemed like a cruel end for someone who lived and worked in some of the most beautiful landscape in the Lake District. He seemed utterly trapped. For some years previously my grandmother had feared he would die in the fields and they wouldn't know where to find him – she would shout angrily at him, 'The crows will go with your eyes.' He would smile, put on his jacket and go back to the fields.

But now there was no going back to the fields.

I am wearing blue suede boots. Don't ask me why, I am seventeen and dumb and trying to be cool. I look like an extra in a Blur video *circa* 1994. I am visiting my grandfather in the hospital after his stroke. He is dribbling out of one side of his mouth and looks like a trapped animal. He is furious with his inability to control his mouth and speak clearly, which makes it worse. One quick glance as I came through the door and I knew he was going to die. Still, he looked pleased to see me, and was amused by my blue suede boots. He can't speak much, but his arm reaches down and points to my feet. A dying man, who can't even speak properly, is teasing me about fashion. When my father comes in, my grandfather clutches his hand and says one almost broken word, the name of his farm. Then he sits and listens keenly to every detail of work taking place on his land, keeping a keen eye on his son and me for any sign that he was being told a good news fairytale for a dying man. My dad and granddad might have spent years fighting, but now they look like best friends. My grandfather is almost tender in a way I have never seen before. He looks scared and keeps looking at me as if to check I believed in all he'd worked for, but he didn't need to worry. I did, and I still do.

When he looks into my face, we share a thousand unspoken thoughts about the farm and our family. In that moment I'm not just a grandson, I am the one that carries on his life's work, I am the thread that goes to the future. He lives in me. His voice. His values. His stories. His farm. These things are carried forwards. I hear his voice in my head when I do work on the farm. Sometimes it stops me doing something foolish, and I pause, and do it the way he would have done it. Everyone knows he was a major ingredient in the making of me, and that I am the going on of him.

It was ever thus.

The summer after my grandfather died, I climbed to the woods high above where we lived and looked down over the Eden Valley, a land where the hay was baled and stacked in countless meadows, and cattle and sheep grazed in thousands of fields. I just sat silently and watched the world go by, with my back to a tree. An old greyish hare hopped up the bank, stopped at my dusty boots and took a long slow look at me, then headed off on his way to wherever. Summer-wild cattle grazed past the little wood, kicking up insects in the golden haze of dusk, oblivious to my presence. As I lay against that smooth old beech tree, the world rolled past me like a dream. A kestrel circled high above the woods, ignoring its ever-hungry offspring mewing from the branches of another beech tree further along from mine. And the whole land was bathed in a warm peach-red August glow. Wood pigeons flapped noisily out of the long sun-bleached grasses where ewes and lambs grazed. Away to the quarry, a couple of roe deer does strayed from the darkness of the plantation to sunbathe and graze contentedly.

A strong dog fox made his way along the shadows cast by the plantation, under a wooden gate and along the fence, until I lost him in the sea of grass. Moments later, he reappeared near where the wood pigeons had been grazing when I last spied them. Pigeons scattered all ways, flapping powerfully away at the grasses and thistles. The dog fox pounced again and again, and then, defeated, he trotted out of the long grass and rolled playfully on the turf. Far below me, the first lights of the village flickered on and the last swallows raced one another across the hillside. I knew the old man had gone and would never come back, and that things would never be the same again. Summer was passing.

Autumn

Unspoilt. Unvisited. Until Thomas West wrote his guidebook, the Lake District was unknown and unloved. No poets came, no tourists toured and the average nymph or shepherd saw nothing worth a second look.

From a modern edition of *A Guide to the Lakes* by Thomas West, edited by Gerard M. F. Hill (2008)

The mountains are as a rule a world apart from civilizations, which are an urban and lowland achievement. Their history is to have none, to remain always on the fringe of the great waves of civilization, even the longest and most persistent...

Fernand Braudel, *The Mediterranean and the Mediterranean World in the Age of Philip II* (1995)

The autumn after my grandfather died, my grandmother gave a silver cup in his memory to the auction mart for the champion tup. I won it with the tup lamb I had told my grandfather about before he died. He was the best we had ever had and was head and shoulders above his peers at the sale. I'd prepared and shown him well, so he was at '12 o'clock' when it mattered. During the judging, I'd stolen the high ground in the middle of the ring to stand my tup on. I'd set him perfect, like a king looking down on the others, an old showing trick. He knew he was the best, and we were simply telling everyone else in case they hadn't noticed. My dad winked at me and smiled when he saw that.

The tup was the champion of the sale and sold for the top price, bought by another respected shepherd to pass on his attributes to the fifty or more ewes he would be mated with.

Everyone had forgotten that I'd failed my GCSEs. I felt as if I'd come out the other side of school and nothing could stop me. I was my grandfather's grandson.

Then it all started to fall apart.

It was a morning full of grey silences and mizzling rain. Dad emerged from the house in his grey suit. My grandfather's will was being read by the solicitors, and my dad, after thirty odd years of hard work, was about to learn his fate. He had worry written all over his face. He left me with a man called John who sometimes worked on our farm and specialized in filthy jokes. John chattered on all morning in the sheep pens,

'Don't worry, son. Your granddad loved this spot, and he thought the sun shone out of your arse!' I turned the words over in my head, and tried to believe them. Granddad had always threatened to amend his will whenever there was a fight. For years, the farm overdraft had grown inexorably, eating up our capital and leaving everyone worried about how it would be sorted out. At times it felt like our strategy was just to work harder, slog it out until things improved. But they didn't.

When Dad returned, he was calm and resigned to what he'd heard. He had poured his whole life into our farm and the end result was that he couldn't keep everything going. Something would have to be sold. Inheritance is often messy and imperfect for family farms. Men like my father spend their whole lives building up the farm and rarely have much other cash. It is common for the son or daughter who takes on the farm to have to sell land or borrow money to pay off sisters and brothers. If you are the one on the farm, it probably always feels rotten.

Over the following months, the bungalow was sold and a flat bought for my grandmother in the local town. There was talk of selling all of my grandfather's farm, and only keeping the rented farm where we lived, but in the end Dad kept the rented farm in the Eden Valley, plus the land from my grandfather's farm. The bungalow and a couple of fields where my grandparents had lived were sold. Granddad's land was to be farmed by us, but remotely now, from fifteen miles away. This had some very real practical repercussions: there was no longer a house for me or my parents to move to in the future. There was nowhere for one of us to go in the coming years. This broke my heart a little bit.

So on that grey day when my grandfather's will was read, my dad didn't want to look at me. He told John what had happened and I listened. Then he turned to me, looked me in the eye and said, 'I'm sorry, son.' And I tried to be a man and smile – a tough smile that wasn't true.

In the months that followed, my dad probably needed me to shut up, be quiet, work hard and support him. Just help him work our way through this tough time. But he didn't get that kind of son. Maybe no one does. After my grandfather's death we all moved up a step in the pecking order.

I heard some old men talking about a young lad recently and they said of him, 'That's the trouble with lads, they think they are men before they are men.' I was like that. By the time I was eighteen, I'd worked part-time for about ten years on the farm, and had done three years full-time since I'd left school. I had a head full of ideas about how I wanted to do things. I thought I knew best. As far as I was concerned, I was a man. I found people of my own age who went to university childish and pointless.

My grandfather, father and I played out the oldest piece of theatre in the history of farming families. My grandfather had been the patriarch, the boss, who had started our branch of the family and our farming business. Our farm was really his farm. Like lots of old farmers, he'd clutched it tightly to his chest when he was old. My father was assigned possibly the worst role in the play, that of suffering the father as boss and the son as usurper. He was doing the lion's share of the work, and never quite getting the control of the farm that his efforts deserved. I was assigned the role of the blue-eyed boy, apple of my grandfather's eye, the perfect farm lad who would be the farm boss some day.

Father. Son. Grandson.

Some fathers and sons we knew seemed to work together like mild-mannered friends. Not in our family. Fathers and sons in our family tend to bicker like hyenas around the remains of a zebra. For a few brief years in my late teens, we would fall out about everything and anything.

The lesson I took from my dad's life was that

if you let your father push you around, you could work for a pittance for maybe twenty years and ultimately not be able to afford to keep the farm together. I wasn't up for years of playing second fiddle, and quite possibly putting myself into the trap he had just escaped. And he'd egged me on to think like that for the past decade or more. But now the wheel had turned and he was the boss and I the son.

Maybe things would have been easier if the farm had been making money. But it wasn't. I was more militant because I could see that I might end up with nothing, even if I served my time. My father couldn't be generous even if he'd wanted to be because there wasn't much to be generous with. So our relationship deteriorated. Eventually it broke. It was at least half my fault. Back then, we were like the meeting of a rock and a hard place. There were only two possible outcomes: buckle down and accept he was the boss; or leave and do something else. My father had left briefly when he was young, and had worked in a local quarry after some bust-up with his father.

As a kid, I didn't see how prescribed these roles were by circumstance. I thought I was special. I thought that there was something wrong with my dad. It was all going wrong on his shift, so it was his fault. My grandfather was the only one to follow and respect. He had created what we had. I look back and realize I was wrong about all of this. I suppose that's what growing up is, realizing how little you know and how many things you've been wrong about.

I look back now, many years later, and laugh at us. We have suffered each other, shared our worst faults, seen each other at our most worn-down, snapped at each other. But I wouldn't change any of it even if I could, because I know my dad, and granddad, in ways that most people never do. I saw and shared in their finest moments. I was part of their world, and understood the things they did and cared about. I let them down at times, as they let me down too. I made them proud at times, as they, too, made me proud. We clashed sometimes. But who wouldn't. Our lives were entwined around something we all cared about more than anything else in the world. The farm.

I am sitting on a hay bale in our barn, I'm four years old. My grandfather sits next to me with a pair of hand shears in one hand and a carding comb in the other. In front of us is a Suffolk tup tied by a bale-string headcollar to a hayrack. It struggled for a few minutes when he first tied it up, but now it stands patiently enjoying being pampered. From time to time it burps and it smells of grass. Two other sheep are tied up on either

145

side of us and my father and mother are working on them. They are cleaning their legs, scrubbing their faces, tidying up their coats and trimming under their bellies with the hand shears to give the sheep clean lines.

Up the village, our relatives and neighbours are all working on their sheep as well. There is a keen sense of competition between us. Shepherds are judged on the quality of their sheep relative to everyone else's. For years, I copied and learned from the older men until I could do lots of the work myself. At the sales, these little things we do can make the difference. The men chat about the best sheep they sold in the past, and whether these are as good or better. I tell them that I like the one we are working on most. My granddad tells the other men I am a good judge and I swell with pride. These sheep are the descendants of two pedigree ewes he bought for a lot of money in the 1940s, and are now part of a flock of sixty. We sell thirty tups each autumn.

Everything that makes us who we are culminates in the autumn. Sheep farms, particularly fell farms, earn most of their annual income in the few weeks of the autumn from September to November. There are literally hundreds of different sales and shows throughout the countryside of northern England. It is about the matching of people with winter grass on lower ground and people with a surplus of sheep produced on the higher ground through the summer. But it is also about more than practicalities: it is the time when we make the decisions that define the quality of our flock. The most prestigious part of these autumn sales is the production, preparation and sale of the tups in each breed.

Improving a flock of sheep is, in theory, simple. You need to buy a tup that brings better genes to your flock: choose him well and he makes your sheep better quality, more beautiful and, ultimately, worth more. The flock of ewes is your core asset, it rolls ever onwards, fixed to your farm, but half of the genetic package each autumn is the tup you buy to match to them, each tup mating with as many as a hundred ewes. So good shepherds are obsessed every year with identifying the tup, or tups, that will have an improving effect on their flocks.

There is a kind of genius to this, in spotting from the hundreds available the one that will match your flock. It matters deeply. The value of your sheep and their reputation can rise or decline rapidly depending on these decisions. A great flock has a particular style and character that reflects the hundreds of judgements that went into creating it, sometimes going back many decades or even centuries. It is not just the sheep that are handed down through the generations, but often the

147

philosophy too: ideas about which characteristics to focus on so as to retain the character
of the flock. Fashions change over time, and
flocks sometimes go out of fashion. Then the
shepherds have to choose whether to change
their approach or hold tight and wait for their
favoured traits to come back into vogue. I
find the depth of commitment and thought
in this whole endeavour breathtaking.

The first tup I ever sold was to a lady called
Jean Wilson. I was nine years old. She was
friends with my grandfather. I was told she
was coming to buy a tup and that my father
would be away working on a piece of distant
land. I was to get the sheep into the yard
with the dogs, show her the ones we were
willing to sell, and to negotiate a sale.

'She's not daft,' I was told. 'She'll be fair
with you but she strikes a hard bargain. Be
ready for her.' Dad told me the price he
wanted – £250 for the best one and less for
the others.

Jean was a born-and-bred sheep woman
and has forgotten more than I know now,
but I'd helped sell sheep for years previously
and knew how it was done. She arrived in
the yard after supper, asked me if I was 'in
charge of sheep-selling operations' and then
smirked when I said I was. She followed me
to the sheep pens.

She pawed over those tups, had their faults figured out in minutes, and interrogated me about which was the best one.

I told her she wanted 'The thick-set one with all the bone. He's the best bred and would do you some good.'

She smiled. I'd grown up with these sheep and knew their breeding inside out. She liked that. 'Aye, that's what I was thinking... But what's he going to cost me?'

'Three hundred.'

We both knew I'd over-cooked this a bit.

'That's far too much. I was thinking £180 was plenty?'

'You can have that little one for that, but not the smart one.'

She didn't reckon much of that idea, as I knew she wouldn't. She was set on buying the better one. So I tried to give the impression we weren't bothered about selling, we could keep it. About an hour later, after we'd danced around the other options, established which school I went to the weather, again explored the prices of the others and dismissed them in turn, we returned to the one she wanted. I told her another man wanted it and he'd not quibble about the price.

She bought him for £250, but demanded and got £10 'luck money' to 'help him be right'. When my dad got home and heard the deal he said, 'Bloody hell, I thought she'd get you down to £200...' Then he laughed.

My life was simple in the years after I left school. I worked. Ate. Slept. Worked. Ate. Slept. I had my evenings mostly free, nothing in them except watching TV with my family. In our house the TV stayed on whatever channel Dad wanted and you watched that; you could sometimes persuade him to change, but mostly it stuck on that channel even when he fell asleep. Some Clint Eastwood movie would be on (Dad loved the one with the orang-utan that punched people when Clint said 'Right turn, Clyde'). If Dad was awake, he'd clap and rub his hands excitedly at the best bits. If he was asleep and you tried to change channel, he'd sit bolt upright.

'Hey, what you doing... I was watching that.'
'You were asleep...'
'No I wasn't, put it back on.'

In another room, my mum, seemingly from another planet, would be ironing or doing some paperwork. She loved Rachmaninov, the Russian composer, and would often play his *Rhapsody on a Theme of Paganini*, either the record, or attempting the real thing on our old piano, sounds of another her that I suspected I didn't know very well.

My father once threw down his cutlery in

the midst of us arguing about something on TV and shouted to my mother: 'I told you. I bloody told you ... you bring them up to have opinions and this is where it ends up. everyone in this house thinks they know best ... even the fucking dog.'

It was beginning to feel as if there were too many people in our house, too many opinions on our farm. I was growing, but it wasn't clear whether there was room to grow. So I escaped into books.

I may never have met my other grandfather, but he also changed my life and shaped how I saw the world. He went to fight in Burma in the Second World War and I inherited from him a nine-inch knife he had taken from the corpse of a Japanese soldier. It was a strangely out-of-place artefact in my chest of drawers where my socks and underpants were kept. My mother had inherited a few dozen books from him and they sat in her bookcase mostly neglected. Penguin paperbacks, with their orange and stained white coats, and their yellowing, long-unthumbed pages, mixed with sun-faded green or brown hardbacks. Legendary books from the 1940s, '50s and '60s by authors I'd not heard of yet: Hemingway, Camus, Salinger, A. J. P. Taylor, Orwell. Turns out my grandfather had impeccable taste in books. And I lucked out because they ended up in front of my hungry eyes at just the moment I needed them.

Each night I would lie awake on my bed, pleased to have the space away from other people, reading like a maniac. When I left school I didn't read much, but very soon I became a devourer of books.

I'd often leave the window wide open, so I could tell what the weather was doing, and hear geese passing over or the friendly chatter of the swallows on the telephone lines. Sometimes I'd climb down from the window, with a book in my pocket, and go for a walk across the fields after they thought I was asleep in bed, and listen to the plain call of the curlews, sounding like the ghosts of dead children.

I'd watch the sun set away to the west.

I'd trek home with the orange lights of the farms twinkling across the fields in the dark, and climb back into my room. I'd be woken the next morning, book still in hand, by jackdaws making their metallic calls as they stole sheep feed from the troughs in the barn below.

One day, I pulled *A Shepherd's Life* by W. H. Hudson from the bookcase as if it was a piece of junk. It was going to be lousy and patronizing, I just knew it. I was going to hate it like the books they'd pushed at us in school. But I was wrong, I didn't hate it. I loved it.

The inscription inside the front cover read 'G. Naylor, Upper V Classical, B. G. S.' My other grandfather had taught at Bury Grammar School: B. G. S. The fact that he'd read books about shepherds made me smile. I opened the book at Chapter IV, 'A Shepherd of the Downs', suffered the first few paragraphs, and then fell under its spell. Two things I loved: the brilliant plain story-telling, no messing about; and the sudden life-changing realization it gave me that we could be in books – great books.

Until I read this book, I thought books were always about other people, other places, other lives. This book, in all its glory, was about us (or at least the old Wiltshire version of us). It is the life story of a shepherd called Caleb Bawcombe, as told to W. H. Hudson when he was an old man in the early years of the last century. I knew the people in this book. They were my grandfather, my father – everyone I knew and respected. I felt as if I could have worked with Caleb and talked sheepdogs, lame sheep, or the weather. There is enough of the old shepherd in that book that I could forget that Hudson's pen was between us. By around midnight I'd read the book and went to find my mother who was still ironing. 'Have you read this? Hell, it's good, Mum... It's about people like us. Have we any more books by this W. H. Hudson?'

We didn't. But her eyes were smiling. Later

I would learn that the shepherding way of life known to Caleb Bawcombe had disappeared, just as, all over the world, the older ways of farming have been swept away by the need for efficiency and scale. Where it does remain is in the mountains, where there is no alternative to the old ways. By the mid-twentieth century, native breeds were sold and replaced by new, improved livestock, and hedges and walls ripped out for bigger fields and machines. Caleb Bawcombe wouldn't recognize farming in Wiltshire now.

When Ernest Hemingway was asked what books an aspiring writer should read, he said you had to read the good stuff by the 'big boys' to see the level you had to beat. He listed W. H. Hudson as one of the benchmarks. Today he's mostly forgotten. But even more than Orwell or Hemingway, W. H. Hudson turned me into a book obsessive, someone who believed in the written word. Suddenly my room started to fill with dozens of books. You could chart how much I was reading by the speed with which we got our friend George the joiner to come and put up new book shelves.

At eighteen years old, for the first time, I could slog it out with my dad at much of the physical work. We travelled to my grandfather's farm (which was now our farm, but without a house) to do the work each day. We were bringing hay bales on a trailer from the field in the valley bottom to the barn. The work was simple, but tough. He parked the tractor and trailer next to the heaps of bales (each one holding seventeen or twenty-two bales depending on the height), then we threw them manually on to the trailer. We took turns on the trailer, stacking the bales the correct way to ensure they would stay on on the bumpy road, with them holding each other in place like the stones in a brick wall, each layer laid on differently to cross the joints. The hardest job was to throw the bales up from the ground, which becomes harder as the trailer load grows upwards. It was one of the muggiest, hottest and most airless days I have ever experienced. We were wet with sweat within half an hour. The bales seemed heavier as the morning went on. But as we filled each load, we got a breather as we drove the tractor and trailer up the winding lane to the field house (one of our names for a stone barn where we stored crop for the winter).

Then, at the barn, my father scurried up the elevator into the hay 'mew' and I pulled the rope that starts the engine-driven elevator, and away it turned, rattling and spewing out petrol fumes to add to the sweat and the dust of the hay. I lowered the bales on to the elevator from the trailer and they departed into the darkness of the old barn, then another bale and ever onwards. At midday we both remarked on how little air there was; we knew that after our picnic and bottles of orange squash were gone, we would be out of drink. But after half an hour lying in the shade we recovered and went back for the other half of the bales in the hay field. We got ever sweatier and more fatigued. Our drinks ran out about an hour after lunch and we both started to worry about how we would be able to keep working for hours without it.

Usually we drank from the becks (streams) or a water trough, but these were almost dried up, and covered in an unattractive film of dust and flies. Our neighbours were all away from home, and the nearest shop was half an hour's drive, away, almost halfway home, and neither of us had any money on us to buy anything anyway. We could have packed up and gone home, but that was fifteen miles away and the sky was darkening away to the west as if it was soon going to thunder. There was little choice but to see this thing through, get the hay in the barn, and then head home.

At last we had just one load of bales left to load. Each bale was weighing like lead and some of them were tumbling back off the trailer (where they will need to be picked up and thrown back on again) instead of floating up as they had done earlier in the day. As the final bale went on to the trailer, we both knew that we were dehydrated. But as we reached the old barn to complete our work, the trailer wheel hit a stone and the whole load tipped off on the hillside, and some of the bales tumbled down the bank. My father and I exchanged a look of absolute horror, and we both smiled darkly at a day that was getting worse fast. The bales had to be lugged back up the bank, and then thrown on to the trailer again. Eventually we got to the barn, unloaded the hay and bolted the big wooden doors that mean the hay is safe and dry. As we left, we were both breathing hard and heavy, gasping for air that wasn't there. We stopped at a friend's house and drank a bellyful of cold tap water, and immediately felt sick. We had heatstroke. The next day we both had headaches and were feeling poorly.

My father always reminds me of that punishing day, like another kind of dad might remind another kind of son of a family day out at the seaside. Something he and I had done together.

Later that year, some gawky teenage town kid called my dad a 'sheepshagger' and then told him to 'go fuck himself'. I expected my dad to knock him to the floor, but he didn't.

He'd told the kid he couldn't walk at will over our land, and that by leaving gates open sheep in different flocks had been mixed together.

The kid swaggered off, laughing, and my

dad turned to me and shook his head.

I'd somehow convinced myself that working hard on the family farm would be a thing that people respected, that people would hold me in high regard as I had my grandfather. I soon realized that this wasn't the case. In our family, it was normal to work on the farm, it wasn't news. Outside the family, it was of no significance to anyone. The result was that I felt I had ceased to exist as far as the rest of the world was concerned, that I was sort of swallowed up by the farm. On the one hand, this was fine by me (I was so glad to have escaped from school), but it was also annoying, because it seemed to confirm that if you went to university you somehow mattered, whereas if you worked hard in an old-fashioned, rooted kind of way, you were not worthy of interest or praise. I also couldn't help but notice that certain young ladies in the nightclubs of our local town showed no interest whatsoever in me the minute they established I was a farm worker.

An old shepherd is hunched over a silver aluminium gate. The square pens in front of him are crammed with grey woollen backs. He speaks, but only the sheep listen, and me, because I am passing.
    'These are bloody good well-bred sheep... You young ones should have your arses

kicked, to let them ewes be thrown away like that.'

He is angry because the auction mart has just sold these sheep badly. They were sold for £22 each. A catalogue misprint had them as 'draft ewes' when they should've been advertised as 'stock ewes'. The difference between these two words is huge to this old man. One word – draft' – means they are the spare lower quality; or usually the older ewes in the flock. The other word. – 'stock' – means they are the core ewes of the flock itself, that a flock is being dispersed, ended. He thinks they have been disrespected, ignored and forgotten in the acres of wiry grey sheep. They were balloted early (the sale order is pulled out of a hat to ensure that no one is favoured over anyone else in the sale order). So they had been forgotten, mostly ignored, and sold before a decent crowd could get to the ring. That autumn we saw ewes given away because they were worth so little, and others sold for just a few pounds each.

His words whisper out over the pens, lost in the wind that sweeps over the fields, roads and towns and away across the fells. The day the flock was sold, the auction was half-empty. Cars and wagons streamed along the A66. Not a soul from the 'farm' attended. They were 'at work' in the local town, having ceased to be sheep people a generation before.

The past falters and dies by little steps. Then it has gone, and old men go home disappointed.

I went home that night and wondered if anyone really cared about what we did. I wondered whether I would just end up some sad old man talking to myself about sheep that no one would give a damn about. The 1990s and 2000s were a time when the prevailing thinking was that we, small farmers in marginal areas, were yesterday's people; the future of our landscape would be tourism and wildlife and trees and wilding. Each autumn, more of the great flocks would be sold or reduced as old men retired, or as various environmental schemes resulted in thousands of hefted sheep being sold off the fells to reduce the numbers. They call it 'destocking'. Some of it is necessary to redress the twentieth-century excesses that we are as guilty of as are farmers elsewhere. But it was felt as a grave insult to many folk who work this landscape because the loss of one flock, or its reduction, weakens the whole fell-farming system, making what we do more fragile. When people in positions of authority spoke about our home, it felt as if they valued everything in it except the things we valued, that producing food was a pathetic cheap thing. This wasn't the farming fairy story I'd imagined, with me as the blue-eyed prince.

I'd always known, of course, that farming was a tough way to make money. But now I was aware just how tough things were becoming. It was the same for small farmers everywhere. Our sheep made the same price at market as they had done twenty years earlier. We kept more and more and made less and less money. The cost of everything else was rocketing. Farm workers got old and were never replaced. Our buildings were thirty or forty years old and slowly falling apart, but there was no money to replace them either. Our tractors and machinery were also ageing. Farming was changing too, with a raft of new regulations that would cost a fortune to abide by on an old farm like ours. My father and mother and I were working like dogs, running fast to stand still, but it just got worse and worse. No one in our world knew what to do about it, other than work like hell and hope that something would change. My dad's catchphrase was 'farming is fucked'.

Farmers had once been the 'pillars of the community'. But the kind of people living in the Lake District had changed markedly during my youth. As houses and little farmsteads came up for sale, they were nearly always bought by people from outside. When my grandfather bought our farm in the 1960s, there were as many as twenty-five little farmsteads in our valley. A lot of the people that lived in them earned money from things other than farming, but they were still country people. The new people who moved in seemed to have little or no connection to our way of life and idea of the land. We called them 'incomers' or 'offcomers'. Some of the old folk called them 'foreigners', despite many of them coming no further than from the local town. The difference we felt between 'us' and 'them' was cultural – they were almost all middle-class English people with professional careers.

If you are objective about it, there is undoubtedly an upside to new people coming into a community – they often bring new ideas, new blood, new money, new businesses and an energy that can renew a community. Even something as local as my grandmother's gingerbread owed its ingredients to the Atlantic slave trade economy. And we are all, of course, made up of many 'ingredients'. But, aged twenty, I could only see what was being lost.

Sheep to the 'incomers' were things that held them up on the road or escaped and were found grazing in their gardens. At worst, some of these people had such a strong sense of 'ownership' of our landscape as a kind of public commodity that they believed they should have a powerful voice in its future. If anyone threatened to build anything new, they would launch vociferous letter-writing campaigns and scare the wits out of the planning officers. A neighbour of ours who has lived and worked here for more than fifty years was so baffled by their attempts to stop a ruined farmhouse being restored that he said to me, 'I don't know how the Lake District would have gotten built if these buggers had been around stopping everything.'

People moving into the villages wanted to buy the village green in front of their homes, which was common land, and were angry when they were told they couldn't. If you walked cattle or sheep through the village now (as we had for centuries), you were liable to get aggravation, because they ate the flowers or left hoof marks on the increasingly manicured lawns. Some new neighbours called the police because they could hear men shouting on the fells above with dogs. We were simply gathering our sheep. Two worlds that didn't understand each other were colliding. It felt like we'd missed a meeting, and in our absence someone had changed the rules to make what we did cease to matter.

Shepherds hate other people's dogs near their sheep. Whilst our sheepdogs are our proudest and most loved servants, other people's dogs pose nothing but a problem to all of us, because a dog that hasn't been trained and is left off the lead near sheep can get too excited and go into full hunting mode. It can be difficult to get dog owners to understand the threat, but about once every two years, throughout my adult life, a dog has given chase to a sheep or a lamb, and before you know it the sheep is pulled down, or lies down, exhausted. Because the dog has not been trained to control itself in these situations, it tears at the wool or the skin, until you have a sheep with its ears torn, or its throat ripped out. About once every other month we have a dog incident that threatens to escalate.

Frankly, every time I see a dog off a lead I am left fretting and fearing the worst until I see it back on the lead and packed into its owner's car. So I suffer, even on the nine out of ten occasions when it all ends OK. Paranoid, maybe, but it is my job to fret about my sheep's safety, and responsible visitors know that this fear exists and act accordingly. One person's freedom is another man's misery.

The ethics of handling dogs and sheep here are simple, and have been forever. Keep your dog away from other people's sheep and it is none of our business. Let it loose, so it begins to chase them or worse and you have put your dog in the path of a bullet. Shepherds have always had a legal right to protect their flocks. Faced with a rogue dog, we have a right, and a kind of duty, to shoot it. If we call the police to intervene, they may tell you to shoot the dog.

Two years ago I was shepherding my pregnant ewes and noticed some sheep whirling in a flock where the fell met the clouds and mist. Something was wrong. I caught a yapping bark or two on the wind. I ran up the fell-side. There was no one else around to sort this out. It was a grey, windy and wet day, and I was running with my plastic waterproofs on. Shuffling along. As I reached the lower fell where there is woodland, a sheep crashed through the undergrowth towards me and lay at my feet. A few yards after it came two Jack Russell dogs. Tiny dogs, but their blood was up. They barely noticed me and dived on the ears of the ewe, tearing them and covering her with blood. I yanked them off by the scruffs of their necks. They were all teeth. I was furious. God only knew what carnage was up the fell or how long this had gone on for. The dogs were maybe a tenth of the size of the ewe, but that made little difference. Even a strong ewe will tire eventually and lie down if the stress overloads her system.

As I pulled the snarling dogs away, a man tumbled down through the trees, full of apologies and mortified at the situation. He'd let them off on what he thought was an empty mountain. They had begun to hunt the sheep until they were oblivious to his calls. But I had a dilemma in my hands; two dilemmas. To me, he had shown that he could not protect our livestock from his dogs. If I handed them back, he might lose them again immediately. I had a mind to have them killed. The man took full responsibility and admitted it was his fault, not the dogs'. But I told him that went without saying. He started crying and begging for their lives, and clutching my arms, which only made me angrier. I had blood on my hands and didn't know whether it was mine or the sheep's. The truth is, I considered smashing their heads against a rock, and I was furious enough to have done it.

But the ewe wasn't mine, so I told him instead that I would take the dogs and hand them over to the owner of the sheep, and the shepherd (a good friend of ours) could decide on the appropriate course of action. I told the dog owner that I would report him to the police and he could face the consequences.

Later in the day, when I had calmed down, and was feeling guilty for being so tough on him, I learnt that the farmer had handed the dogs back after reading the owner the riot act. The police had explained the seriousness of the situation and that he was lucky to get them back alive.

When I was growing up, I thought my father and his peers were needlessly rough in their handling of situations like that. They would swear and give people an old-fashioned rollicking if they had their dogs off leads where they shouldn't have. I used to think that some gentle explaining would suffice. But you live and learn. I've tried the gentle friendly approach until I am blue in the face and it generally results in the other person carrying on regardless. So now I rant and rave and generally do a great impression of a furious farmer who will shoot their dogs with immediate effect unless they are put on a lead.

It works much better.

Most of my memories of those years after I left school are of working with Dad through the changing seasons.

An eel slithers across the mud back to the water. With each mechanical handful of sediment that he scoops out of the channel come dozens of eels. He is sitting on an old faded yellow Ford digger. I am holding a spade, and jumping about gathering up eels in my hands. Most are little longer than shoelaces, and hardly as thick as a pencil. They have made their way here from up the rivers and from across the seas (the Sargasso Sea, the books said, but I had no idea where that was). They have buried themselves deep in the mud at the bottom of our little becks, but not deep enough now.

Sometimes a much larger eel would writhe out of the disturbed mud. My dad shouts, 'Look at that big black bugger. Grab it.'

It is three feet long. I step back, scared by the dead grey eyes and its muscular snake-like movement. It writhes back into the muddy water. Countless little becks and drainage channels like this let the water run away. My father and I are cleaning out the small stream that runs through our land and drains our hay meadows, lowering the water table and keeping our meadows as grass instead of slowly becoming a field of sieves. But the streams inevitably silt and stone up within a few short years. If we want to maintain our land, we have to work at it – it is not naturally like this. The old men know where every drain was hidden on the land, and pride themselves on it. As the streams are cleaned out, the terracotta drains reveal themselves and start to trickle again. Sometimes we pull out lengths of hollowed tree trunks, split down the middle that have been laid upside down to create a drain; at other times, the drains we unearth are handmade with rough stone, from many centuries ago. There are archaeological remains in the neighbouring valley of some of the small farms scattered across the valley side three thousand years ago.

My friends and I used to hang out in a pub

a couple of miles away from my home. We would do dumb stuff for kicks. Once, someone came into our local and said there was snow on the fells about fifteen miles away. So we got in a car and went and filled the boot with snow. When the local nightclub closed, we bombarded people with snowballs – which fairly baffled them for about a minute because it was quite mild, with no snow at all in the town. Then the mood changed and we had to leg it from a bunch of big lads, who were intent on giving us a good kicking and chased us off through the streets. Running for our lives. I could hear the Iggy Pop song from *Trainspotting* in my head as I raced down the little lanes between the houses, some big lad's feet thumping the tarmac behind me. They didn't catch us.

Everyone always seemed to be fighting or screwing someone they shouldn't have. One of my best mates used to get so drunk he couldn't stand up, and we'd have to take him to A&E where the nurse would groan at the sight of him. About once a month he'd smash his face up falling over a wall or something.

I remember one night we got into a fight with some lads, and the next day the police called and said one of them had been found dead. After fighting with us he'd got into a fight with some other lads and that was the end of him. The police knew it wasn't us, but we had to make statements. It was that kind of northern town.

One Saturday night I came out of the chippie and a lad I knew was punching someone who seemed to be unconscious in a side alley where we stood to eat. We pulled him off, and the other man got up, groggy, and went home. I remember feeling sick. Something could easily go wrong and then one of us would be caught, banged up and the game would be over. Fighting wasn't really my scene.

Drinking. Fighting. Shagging. I could see the future spread out in front of me and I didn't fancy it. But it was hard to know what to do; the optimism I'd had after leaving school had drifted away.

Then one night we were at a party and I met a girl called Helen. She had red hair and was pretty. She was a friend of my sister. I was twenty-one; she was eighteen. She had worked hard at school, read books and knew all the stuff I didn't. She was smart, and confident about it, and puzzled that I wasn't. When I was around her, I was just able to be me, and I began to feel tired of being anything else. She believed I could do anything I set my mind to. That made everything possible.

From the moment we got together twenty years ago, she made me want to buckle down and make our life a good one. She makes me better than I am.

My friends were horrified at the sudden change in me. A good fool ruined. If she was out in the pubs of our town, I would find a corner to talk to her and I couldn't be moved. No more clowning around. Instead of entertaining everyone with mischief, I had tuned out.

In our local pub there were shelves either side of the fireplace lined with books that no one read, just there as book wallpaper. Occasionally I would take one if it caught my eye. Quietly I'd ask the landlord if I could borrow it and stuff it in my jacket before anyone else noticed. It wasn't cool to be into books.

One of the regulars was a veteran of the Korean War. He'd mown down thousands of Chinese soldiers as part of a machine-gun crew, as they had yelled and charged at him from across a valley. They'd mowed them all down and the valley was full of corpses and screaming men all night: they had lived those dark hours in a human slaughterhouse. His hands shook a little as he told these stories.

One day I took a book off the shelves in the pub, a memoir of the Second World War, written by a German fighter pilot, who, strangely, had had no regrets about fighting for 'the Führer'. The Korean War veteran saw and asked what it was. He looked at it and then declared that we youngsters knew nothing, we didn't even know what the plane on the front of the book was. My mates looked blank. He was about to set us straight.

'Messerschmitt 109,' I said.
'What?'
'Messerschmitt 109 ... a G2/R2, I think.'

Silence. Everyone was looking at me weirdly, wondering if I had made it up. Then they all looked at him. The old veteran nodded, then started to grin.

Weeks later, there was a quiz in our local pub. We didn't usually enter a team against the teams of teachers and professionals who turned out and appeared to know bloody everything. Instead, we'd sit and drink and wind each other up, or play pool. Then one of my mates figured we could win a quiz about the Second World War. About two hours later, we were winning the quiz, and I was pretty much the only one answering questions. My mates were grinning from ear to ear, and slating teachers at the bar about how easy it was. We were getting funny looks from the other teams, who were trying to work out how the 'village idiots' could be winning the pub quiz. In the end we lost by a point on the last General Knowledge round, a question about a 1960s TV show I hadn't even heard of

Later that night one of my friends said,

'What are you doing here ... with us fucking idiots? You should go to university and do something smart! You're smarter than these teachers. You should fuck off and do something.'

I was unsettled, because I didn't want to be any different from anyone else. And I certainly didn't think I was any better or smarter than them. It was a stunt, some showing off based on having read a few dozen books about the war. But sometimes you can't go back when people know something new about you.

Things got bad. My dad and I fell out about a tup he'd bought. I hated it. With hindsight, it was probably not a bad specimen, but at the time it wasn't what I thought we needed for our flock. It had some black wool in its neck that it shouldn't have had. He thought it was a birthmark and wouldn't be passed on to its lambs. I disagreed and thought it would put a horrible flaw in our flock for years to come. It would have been easy to try it on a few ewes and see, but it provided a spark for us to explode at each other. For weeks we fought, bullied, said cruel things, tested each other, spotlighted each other's flaws and scored points in front of others. There were times when I could have killed him, and I'm sure he felt the same. On odd occasions we were dragged off each other, fists flying.

I said I was off, that I was leaving the farm. I knew I could swallow my pride and back down, but it wouldn't solve the big issues and I wouldn't respect myself. I'd watched people who should have left their farm and done something else, but had stayed too long. You could see them becoming surly and bitter, and I could feel that growing in me too. But I didn't really know what to do. I barely knew what a CV was. If I'd had one, it would have read: GCSEs – didn't try and failed. Work – on a farm. And I had no money at all. I didn't even have a car and I lived fifteen minutes' drive from the nearest town. I guess the best thing I had going for me was that I had nothing to lose.

My two younger sisters turned out to be far smarter than I was: straight A-grade students, the kind of girls that get on the front of the local newspaper, clutching a piece of paper after their exam results. Sometimes I'd help the elder one with her homework. She found it funny that her brother who had flunked school read so many books and knew so much stuff. One night she challenged me to do her history homework. I think she had a hot date or something, so she left me to do it. It was like a joke between us to see how I did. I stayed up late and typed out (with one finger on a word processor she used) the essay. A few days later, she was seriously pissed off because the essay came back with a rave review from her teacher. He told her it was even better than her usual high standard. I laughed, and told her that school was a 'piece of piss'. She told me that was it, no more goes at her schoolwork. She made it clear that whilst I might be brainy, she at least was at school and would get A-levels – I wouldn't. So from that moment on, I kind of knew I could get A-levels if I wanted to.

My younger sisters were never as indoctrinated with the farm as I was. They were part of it too, but not blinded by it. They were always a lot more modern and 'with it'. This was partly because attitudes were changing rapidly around us. The four and eight years respectively that separated me from my sisters made a big difference. It was also different because they were girls. In many farming families, the daughters have none of the boys' embarrassment about being different, and know that their role is to leave and do something else to earn respect (either that or marry another farmer and start their own branch of the family doing much the same as their own folk). The same farmers that are proud of their sons for flunking school to come home to work on the farm are also proud that their daughters work hard at school and go off to do other things in the wider world. Attitudes to education have changed as well, with many farmers now proud (and perhaps a bit relieved) when their children choose to stay at school and live a different life, but that wasn't so true when I was young. It meant that my mother got more of her way with my sisters, and they repaid her efforts with the kind of exam results to make any parents proud. They got into the local grammar school, and had quite a different school experience from mine. I'd been a little shit, and they were top-of-the-class material. It was hard to believe we were from the same family.

I applied to the local Adult Education Centre when I was twenty-one (it always sounded to me as if they should be teaching porn) to study for A-levels in the evenings over two years. The teacher called me and told me I couldn't get on to the course because I had no GCSEs. I'd have to do them first, then reapply. I asked him whether the class was full, and it wasn't. So I asked him to give me a three-week trial, and if I was out of my depth and a nuisance after three weeks, I'd leave of my own volition and they could keep my course fee for the year. He prevaricated a bit and said it was irregular but he'd go along with it. So after I'd got my farm work done, I'd jump in my parents' car and drive to Carlisle (half an hour away) and sit in classes from 7 p.m. to 9 p.m.

The first week I was very nervous. I told myself I was in control and could leave if I didn't like it. I didn't tell anyone except my family, and asked them to keep quiet about it. I wasn't doing it for anyone except myself, to show myself I could do it. The teacher had offered me the perfect challenge: I hate being told I can't do something. There were about twenty of us in the classroom: a couple of elderly folk studying for a hobby, a couple of youngsters trying to improve their CVs, and about fifteen single mums. There was at the time some weird policy that said you either had to try and get a job or be in education in order to qualify for benefits. So it was common practice for single mums on benefits to do a night a week at college with their friends. They were young, bubbly, chatty, and probably quite bright if they wanted to be, but mostly they were just totally disinterested in the course we were on. I sat with them and enjoyed the fun they were having.

But since I'd left school, something had changed in me. I was on a mission this time, in the room because I chose to be there. It made all the difference in the world. The teacher would ask a question. Silence. The ladies at the back would pay no attention at all. The earnest few would try and answer and get it wrong. And then I'd answer it correctly. It was fairly easy if you'd read the books I had. In fact, after a couple of lessons, it felt a bit like I knew more about some of the subjects than the teacher. I knew the academic debates about different issues. The teacher, to his credit, encouraged it. When I asked him after three weeks whether I was still on the course, he told me not to be so bloody stupid: I'd got straight As. He started questioning me each week after class. Why had I got no GCSEs? What was I reading? What did I do? Did I want to go to university? Had I thought about applying for Oxford or Cambridge? I laughed the last one off, 'No way, I hate students.' And I did.

I never once wished I'd gone to university. The few people I knew who had been didn't seem to have come back any wiser. They seemed to have returned home full of nonsense. And they never really fitted in again. Still, the question threw me a bit, because although I genuinely didn't want to go anywhere, if he thought I was that good at this book stuff, then perhaps it meant I had options. And I needed options. Books and doing my A-levels became a kind of escapism, something that I could control. I was discovering something about the wider world – that you could shape your own fate to a much greater degree than I'd ever experienced. If you read more, worked harder, thought things through smartly, or wrote or argued better than other people, you won. For a while I found this newfound freedom quite exciting and liberating. I found it a bit of a buzz just to be good at something, something that was nothing to do with my family or our farm, or anyone else except me.

I just had one small problem. I couldn't write by hand.

I was always poor at handwriting at school, and for nine years after leaving school had not needed to write anything. What little writing I did – recording sheep numbers and other functional short notes – I did in block capitals. So when I signed up to do my A-levels at evening classes, I wrote the weekly essays on a word processor, typing with one finger and handing in tidy, well-presented sheets of A4. And then whack! It struck me in the face. It was about a month before the exam date that I realized I would have to write the answers in essay form by hand and not in block capitals. Helen set me a pre-exam test to see whether I could write the essay answers by hand.

Half an hour later, I threw down the paper and stormed out of the room. My writing was almost illegible. What was worse, I found it so difficult to hold the pen that I spent all my time focusing on my hand-writing and none thinking about what I was writing. Scariest of all was that the harder I tried and the firmer I gripped the pencil, the worse things got, until my hand was cramping, I was sweating and losing my head. I was panicking – and I was also ashamed. What sort of idiot knows all the answers but can't write them down?

Helen bought me a children's book for learning handwriting. I was surly and ungrateful, but eventually I could scribble words that someone could read, though even to this day I break out in a cold sweat if I am ever asked to write anything more than half a dozen words by hand.

In the years since I had left school I had observed the professional people who were buying houses in our village. These people seemed to earn more money in a week than I did in months, and it looked as if you had to get an education to play their game. So I decided to play along. I was going to whore myself in a world I didn't like. And I figured that if you are going to be a whore, you should be a high-class one. I decided to do something I didn't really want to do. I would apply to university and see if I could get into Oxford. If I could, I would consider going. If I couldn't, I would bin the whole idea. After a few months, I had one A-level to show for my efforts, was halfway through some others, and had a teacher who was prepared to write a gushing letter, saying I was some kind of dysfunctional genius that deserved a second chance.

I'd never been to Oxford, or anywhere like it. It seemed quite ridiculous that I might get in.

But the interview went like a dream. I found myself in front of a bunch of bored professors who did a kind of bad-cop-good-cop routine on me. Aged eighteen, I would have wilted, but I was now in my early twenties and it was easy if you weren't really bothered. So, much to the amusement of the other professors, I got into a row with one of them. I like arguing. I'm good at it. When he went too far, and said something a bit silly, I teased him and said he was losing his grip. As I left the room after my time was up, I smiled at them as if to say 'Fuck you, I could do that all day'.

They all smiled back. I knew I was in.

We were working in the sheep pens, sorting lambs for the auction. We handled each one across the back with the thumb and first finger, feeling the fat covering on the back, pressing into the wool. There is an art to selecting them in their prime. We pulled those that are meaty into a side pen.

I asked my father if I should go. He told me I had to. That they would do fine at home without me. They might even do better, he adds, smiling. I can always come back, he says.

This thing that happened seemed to have

cleared the bad blood between us. Everything was suddenly calmer. The week before, we had been at each other's throats, and now he was my dad again. I was no threat to him and he seemed to understand that I was about to sail into uncharted waters. I think he felt guilty, too, sensing that I didn't really want to go anywhere and that it was because we couldn't get on anymore that I was going. He came home from selling the lambs that night and seemed to be in high spirits. Word had got out about it, and his friends were teasing him, saying I must have got my brains from my mother.

I was briefly a minor 'working-class hero' in the pubs of our local town. When I told my friend David, whose family farmed up the village from us, he looked at me as if I had gone mad and replied with complete sincerity that they must have made a clerical mistake, because I was just an idiot like him. People that remembered me at school couldn't quite get their heads around it. Middle-class girls who'd previously thought I was not quite a full shilling, now suddenly showed a lot more interest. I laughed with my friends about this and one of them said I should just take whatever I could. I laughed. But the truth was I'd already found the girl I wanted to be with.

About ten days into my time in Oxford, I

became aware of how people communicated. Notes were posted in the porter's lodge in a pigeonhole with your name on it. Mine was now jammed full of increasingly frustrated notes from the History tutors, asking why I wasn't attending all of the start-up meetings. Christ. Damn it. I'd thought it had been kind of laidback and quiet. The last note basically said if I didn't come and see them ASAP they would have to assume I wasn't on the course and take action. I'd apparently missed social drinks, the inception meetings, explanations of the library system, and a bunch of other stuff that would have helped. So I went to see the professor and confessed the truth. He looked a bit annoyed that anyone could be that dumb, but told me to go to the library and get on with an essay due in a couple of days. I hurried off to the Bodleian Library where it turned out the books I needed had already been taken by the other students.

The basic drill was that each week you had one or two tutorials where you went face to face with a professor. Two days earlier, you'd handed in an essay on the subject given the week before. You were given a reading list that filled a side of A4. It must have had twenty or more books on it. Your job was to read those books, and others that you might find of relevance, digest them, and then write an essay showing your dazzling originality and clear analysis of the issues. When I was handed my first reading list, I asked the professor how it was possible to read twenty books in a week and do all that. He gave me a look that said 'Just do it'. After a while it became second nature either to actually do all that work, or to do enough so I didn't get an intellectual kicking the next week. The first three weeks I got the second mark from the top, a 2:1, which in my understanding was like a B grade. I asked the professor why I wasn't getting top marks. He told me the marks were good, but I should be more me and less a copy of everyone else there. It hadn't occurred to me that being 'me' was a potential advantage. And then the penny dropped. Everyone in Oxford was bored with perfect kids from perfect schools. Being a bit northern and weird was my greatest strength. It could make me interesting. I could beat the perfect people by doing things they couldn't do.

I was seated next to a professor. He asked me about my life at home, and I told him. I always gave them what they seemed to want, a little mythical version of me. On the application form, I had told them I was a 'drystone wall builder on the Lake District fells', so every conversation for the next three years was about how it must be a big change being in Oxford after working on the fells each day. I had a good evening talking to him and at the end he said, 'Oh well, I imagine you will miss it.' I told him that I hadn't stopped doing it, that I was going back. He seemed quite confused by this.

He asked what I made of the other students, so I told him. They were OK, but they were all very similar; they struggled to have different opinions because they'd never failed at anything or been nobodies, and they thought they would always win. But this isn't most people's experience of life. He asked me what could be done about it. I told him the answer was to send them all out for a year to do some dead-end job like working in a chicken-processing plant or spreading muck with a tractor. It would do more good than a gap year in Peru. He laughed and thought this tremendously witty. It wasn't meant to be funny.

Being in Oxford was strange for me at first. I'd never had a life where empty days stretched out before me. I'd never woken up and wondered what I'd do that day. And suddenly I was waking up, my body clock telling me it was time to start work, but no person or animals needed me. I felt like an island in a sea of other people. I didn't really like the freedom I was experiencing – it felt pointless and empty. I ended up like one of those imprisoned tigers that walk endlessly from one side of their cage to the other. I threw on my clothes each morning and went 'pretend-shepherding', walking for miles around the parks in Oxford, across the fields, past the paddocks full of ponies. I did a lap or two of the quadrant in college – it was as if my body and my mind needed to treat this like my farm and my farm work. By about 8 a.m. I was walked out and no one else had surfaced yet in the college. I called home, but that just annoyed everyone there because they were working and hadn't got time to chat.

They don't say it, but I knew they had to work harder (particularly my mother) because I wasn't there and that made me feel ashamed. I decided that I had to come out of this place with top marks, or else I would have let everyone down. So I started going to the libraries early each morning and worked until they closed at night. I tried to work out how I would explain the subject to my dad or my mates if they asked me to.

But occasionally the sunlight through the library windows would catch my eye, and I knew I should be out in that. I felt as if I had cut myself adrift from everything I loved.

After the first year, Helen came down to Oxford. She baked cakes and sold them in a café and at a farm shop to support us. She put me to work as her baking assistant. The cakes had to be delivered first thing in the morning when the café and farm shop opened, so we baked last thing the night before. We had just one problem; the kitchen in the small flat we rented from my college was tiny, maybe six feet by four feet in total. So the whole flat was taken over with trays of lemon drizzle, chocolate brownies, flapjacks and Victoria sponges. Helen was a good cook, but my skills extended simply to washing up, mixing ingredients to her instructions and carrying things to the car. To begin with, the cakes were a sideline, but after a few weeks the orders got bigger and bigger and we had to work half the night to meet them. The table and sofas were laden with cardboard boxes full of tray bakes, coffee cakes balanced in gaps in the bookshelves, and even the bed had a few coffee cakes plated on it.

We had tremendous rows about my inability to follow instructions, about things getting burnt in the pathetic little oven, or occasionally me dropping something as I went up the flights of stairs from the basement flat to the car. But we did it together, and now, years later, we smile at the silly things we did to make enough money to get by. I think it was a good time. It meant we built the foundations of our life together without too many other people or agendas getting in the way. Later, when we married and went home to the farm and had children, there would, of course, be many distractions, our life would be less about us, and more about the things and people around us. But it would be built on strong foundations.

The strangest thing about leaving the farm and starting to live a different life was that, from the minute I left, I was always coming back. I quickly realized that my new life often left me with loads of time of my own – weekends, holidays and evenings. You often don't need to be physically at a university or an office all the time. The three terms at Oxford were eight weeks long, twenty-four weeks in total. It soon became clear that I could still be at home for more than half the year. I could even be at home for half the week sometimes, during term-time. I wouldn't make many new friends at Oxford and I would remain distant from most of the other students, but I didn't lose any sleep about that.

My boots are covered in sawdust. I am standing in a pen, jostling with other shepherds, where the ewes are held momentarily before they go into the ring to be sold. It is where we can inspect them for the last few seconds. As one lot is let into the ring, my attention either follows them, if I am interested in buying them, or turns down the alleyway for the next consignments coming towards me. I got home from Oxford late the night before, after being there for about a month. It feels strange to be home, as if I am now just a visitor to the land that I love, no longer really part of it. I understand for the first time that our sense of belonging is all about participation. We belong because we are part of the work of this place. So I got up early and shepherded, and half an hour of work makes me feel part of it again, like I can shed my other skin. There was a heavy dew, or 'rime', on the grass, and the ewes' backs were silver. My boots were sodden when I came back in for breakfast. We then drove up the Eden Valley through an autumnal landscape, through air with a bite to it. Sunlight lay like smoke in the hollows, resting, before it made the long afternoon trek up the fell sides. The lichened stones shone silver in the thinning light. The hedges flecked with the blood red spots of rosehips. The chimneys of the farmhouses marked again by first whispers of wood smoke.

My heart aches because I know that in my new life I am divorced from the changing of days and the seasons. Things have changed a great deal in the month I have been away. I see big changes instead of the little ones I have always lived with. Autumn comes quickly here. The life bleaches out of the leaves and grass with each passing day. A landscape of green turns brown. The heather on the fells turns until it is the russet of a kestrel's wing.

As the ewes come down the alley to the bustling ring in their different consignments, interest in them grows until, in the last section of the alley where I stand, we inspect them. At this sale we buy the 'draft' ewes from the fell farms, the ones we breed the hybrid 'mule' lambs from to sell. My father (like my grandfather before him) travels to the sales at little auction marts like Middleton-in-Teesdale and Kirkby Stephen. We go to buy ewes that have lived on the Pennines, but which are now sold for a new life at a lower, less harsh, altitude. The fields and streets around the auctions fill up with badly parked Land-Rovers and wagons. You see as many as three generations of the same families trooping down the streets to the auction. Little fell-bred old men with bent backs and bandy legs, flanked by strapping beefy grandsons about two feet taller. Traditionally, it was a day to wear your good clothes. My grandfather would cast an eye up and down me to make sure I was properly turned out. He'd have on a tweed suit and a tie. Boots polished. I could get away with jeans if I had a shirt and tie on under a jumper.

I feel the sheep's backs for condition, decide their quality with a glance at their colour, fleeces, legs and heads, and check their teeth by grabbing a sheep and peeling back its bottom lip (sheep only have teeth on the bottom jaw). The teeth tell me a lot.

A lamb has baby teeth, little, sharp needle-like teeth, but then at a year old the two central teeth change into broader white teeth. A year later, the next one on either side of the central two change to adult teeth, and the year after that the whole mouth is in its adult form, like a tight little row of tomb-stones made of white Portland stone meet-ing at the edges. As the sheep ages, the teeth get longer and start to weaken, with gaps between them, until eventually they become wobbly and fall out. They can actually graze with no teeth at all, but there comes a stage when the mouth is completely 'broken' and they struggle and lose condition. When ewes are 'broken-mouthed', they are sold for meat because their ability to feed them-selves and produce lambs has gone.

On days like today, my job is to stand and check the mouths of the ewes. I was taught to do this by my father over many years, until he decided I was a good judge. Now he takes my word for it, and sits across the ring from where he can see the sheep and bid if he wants to. Because of their age, these ewes' mouths will be mature, but the art is to judge from the teeth whether they will last several years, or just a year or so. Our judgement of their value is in many ways an assessment of their age and durability largely based on their teeth. A 'good mouth' might mean you get three years from the ewes; a 'bad mouth' could mean just one year. I check hundreds through the day, and Dad watches from the other side of the ring, prompting me with a nod to indicate whether the sheep in the ring have good mouths or are to be avoided. A little smile or a wink and he will know these sheep will wear well. He buys some of them if they are at the right price. A tiny shake of the head or turning away tells him to leave well alone. The difference in price might be as much as £20 per ewe, and a large farm might sell hundreds of draft ewes each autumn, so little things like teeth matter.

A lot of my farming friends see me at the sales, and have no idea that I am now at university. I don't tell them. Others know, and are watching to see if I have lost the plot. One or two folk are not sure what I am anymore. They start to say, 'I thought you were...', then realize it is still me and we talk sheep.

The draft ewes from the most prestigious flocks are coveted because, although they are not young, they represent an opportunity to breed exceptional offspring. They may only live for two, three, maybe as many as five or six more years, but in that time they may breed lambs that are better than your own ewes are capable of breeding. For all these reasons, anyone starting to breed fell sheep seriously has a limited chance to get amongst the best bloodlines. Sales of genuine 'stock ewes' in their prime from a respected flock are few and far between.

So this sale is of generally run-of-the-mill sheep that are effectively commodities: many are bought to breed lambs for meat, but there are also a good number of higher quality sheep that are the object of keen competition and immense pride. The sheep being sold are actually the oldest ladies in the fell flocks (albeit with some miles left in the tank) or those sorted out because they don't breed as well. A fell flock is like a conveyor belt with the oldest (five to six years old) ewes moving off each autumn at the top, and new younger home-bred females (in their second year of age) pushing on to the bottom to take their place. Every year the flock is renewed by fresh young ewes, and through the sale of the older ewes.

One of the stranger features of a fell farm is that your best females are rarely ever seen by anyone but you and your neighbouring fell shepherds.

Shepherds compete with these draft ewes at the autumn sales for the prestige of securing the highest prices, because although they are not our best ewes in their prime, they still have significant sale value and form a big chunk of our annual income, and they make a very public statement about the quality of our flocks. If they have poor teeth, or are old and lean, then the quality of our breeding is thrown into doubt. If they are still great ewes, with good teeth, and have 'worn well', then our breeding looks desirable to other shepherds. Some of the old shepherds say that you can only really judge how effective a tup is when his daughters are sold as drafts maybe six or seven years after he is bought.

Men and women crowd five-deep to peer through to the sheep in the sale ring. When a notable flock comes to be sold, the crowds follow them. Our friends, the Lightfoot family, are heading to the ring with their ten best ewes, their heads and legs sparkling black and white against their fleeces. The Lightfoots are highly respected shepherds with sheep that have a depth of breeding that everyone here knows about. The names of the great flocks are uttered reverentially. Such sheep are presented and treated with great respect. Hours of preparation have been invested in getting them into prime condition. They have been dipped and coloured with peat, their black and white noses and legs washed, and stray white hairs or black ones 'tonsed' out with tweezers. The great-looking ones with a track record might make £300 each, the run-of-the-mill ones £100. The best ewes are even spoken of affectionately: 'She's special that old lass, bred me a tup I sold last year for £3,000.' These ewes can never be put to the fell again, because they are hefted to another piece of land, so they are farmed with care on the lower fenced ground.

My father is wielding a white-handled meat saw usually used by a butcher to cut through bones. I am holding a very-much-alive tup in the corner of the sheep pens, his backside in the corner. His head is twisted strongly up around my body to present his horn at an angle my father can cut it, my knee is in his chest. The tup is angry and throws us forward a few inches with each thrust of his body. I counter it to keep him as still as I can. With each wriggle, my father curses.

Swaledale tups have curled, ammonite-like horns, curling once or even twice round before they are old. In the autumn, when their blood is up, they back away from each other, then charge with their heads down, until they meet with a vicious crack that is like two great rocks smacking together. Then, one might be found lying peaceful, still and ended, its neck broken. Most of these tups have horns that curl safely away from their eyes or from their heads, and can be left alone. But others have horns that break, or twist the wrong way, or grow into the tup's head, or grow so fast that the flies aggravate the base of the growing horns.

So we 'train' the horns of some sheep, and have to be vigilant about others. Sometimes we have to cut off a sliver of horn nearest the head to stop it boring into the flesh. Occasionally we 'warm' a horn to bend it to a more harmless course. We have even used a contraption that bends the horns slowly away from the head by tightening a bolt a little bit each day to apply slow pressure to a chain between the horns until they hold right naturally. If the horn is just too tight to the head, we will take them off as the lesser of two evils. Sometimes there is some blood, but it soon stops, and the end result is safer for the sheep.

When the tups are old or dead, the horns are sawn off to make the varnished soft curved handles of the shepherds' crooks and attached to a staff of hazel with a seamless joint. Nothing was wasted in the old days. Some of the old shepherds or men in the villages carve ornate sheep or sheepdog heads in to these horn handles to decorate their crooks. The best of these sticks are never used for work, and are simply for show. I will wave my crook to get the tup's attention in the sale ring, and tickle it gently under its nose to get it to raise its head to look prouder and full of character.

A crook is as essential now on our farm as it ever was. My crook is an extension of my arm, letting me catch the sheep. Sheep are faster than a man, but will let you within a distance they feel safe at. The crook is used to take advantage of that and snag them around the neck. I use a crook almost every day in winter and dozens of times a day in the spring when we are lambing and need to catch ewes at regular intervals. We also carry with us a medicine box with all the tools and potions that our trade demands – penicillin, purple foot spray, foot shears, multi-vitamins, hand shears, needles and syringes, wormer and fly repellent.

My two-year-old son Isaac understands that a stick is part of what makes you a shepherd. He has his own, made for him by a distin-guished shepherd and stick-maker. Each autumn at the sales the auctioneer sells a handful of this shepherd's line crooks. They are keenly bid for. Some have the classic horn-curled crook handle; others are wooden and in different styles. They are beautiful sticks and much admired, and the shepherd that makes them has made and sold many hundreds for charity.

Last year I had to go and see this old shepherd, we had some sheep business to attend to. I loaded my son into his car seat and we drove off across the mountain passes that take you to Langdale. An hour or so later, we pulled into the farmyard and my son woke up. We were ushered into the kitchen of an old-fashioned, but picture-postcard-pretty farmhouse, and then the shepherd showed me his sticks, dozens of them, many carefully laid out for the day when they would be sold, others hanging upside down from the black oak beams so that their varnish would dry right. He told me that his sheepdog is so good on the fell he could send it from the kitchen table to gather the crags above the house and carry on his breakfast whilst it worked in the rocks above.

The old shepherd was proud of his sticks, and rightly so. I asked him if he had learned from his father, and he said, no, he was self-taught. Outside he showed me the workshop in the barn where sticks in various stages of creation are placed. Bundles of hazel or beech rods, tied around the middle with string, rest against the workbench. In the vice was a beautiful stick that he had been working on. He told me that he was having a job to curl the horn the way he wanted, as it had a little twist in it. I told him I liked it just fine, it had character. He said it was mine. And a few weeks later it was put into my hand, beautifully varnished, and less crooked that it had been, because he wanted it to look right. Then he handed me a stick for Isaac, a boy's crook, with a half-curled horn handle, the perfect height for him to stand and lean on.

My mother is sitting on a wooden chair in our barn, slightly hunched over a sheep's face, her glasses perched on her nose as if she is trying to read some tiny writing. A Swaledale sheep is held tight in the crate that holds their heads as we prepare them for the sales (sheep tend to stand still if firmly held under the chin, with a rope behind their heads, to tussle a little and then resign themselves to enduring the work we do on them). In her hands is a pair of tweezers, the kind a lady might use to pluck stray hairs from her eyebrows. She is 'tonsing' a tup, plucking out some of the hairs on his face to make his black and white colouring look even more distinctive.

There is an ideal pattern, colour and style to the face markings of Swaledale sheep that is aspired to and only very rarely achieved. So everyone cheats a little and helps with the definition of the pattern by plucking. In the barns across northern England, everyone who breeds Swaledales is at it. I knew a shepherd who spent more than forty hours once on one sheep, plucking. When he told me, I laughed and said he was crazy. His response was, 'Aye, maybe – but you should have seen her, she was beautiful by the time I'd finished. Won every show that summer.'

The sunlight lights up my mum's hair. She is always the most patient and careful of us, so she got this strange job, my dad losing patience with it after twenty minutes. We have always done these little things, but I appreciate them much more now when I come home from university and see them with fresh eyes. I realize more than ever that these little things make us who we are.

I suddenly found, now that I was a student at Oxford, that I was invited to 'book clubs' in our village and to dinner parties when I came home. People wanted to talk about current affairs with me when I met them on the lane.

Occasionally someone would ignore my friends if we were out and make a beeline for me to talk about something Oxford-related, and they would just smile. I had been reclassified as 'clever' and I was not entirely comfortable with it because it confirmed lots of things I'd always suspected.

The fields are silver-wet with late autumn dew, and where the sheep have run the grass has been shaken back to green. There is a nip in the air. We are working in the sheep pens. My father's dog Mac lies with his head under the wooden gate, keen to be in where we are working. Selecting which of the ewes will go to which of the tups is arguably the most important work of the year for a shepherd, a complicated business of choosing animals with complementary characteristics to try and breed the best sheep you can. So we look at the ewes that fill the pen in front of us, weighing up in our minds which ewes bred best to which tups last autumn, and which ones might breed well to the new tups we have bought this autumn.

The ewes have the tops of their tails sheared a week earlier to make it easier for the ram to get them pregnant (think removing woolly knickers). Then we dip them, worm them and give them a mineral supplement and a dose to prevent liver fluke disease. A pre-natal MOT, if you like. Our job now is to ensure that the ewes are mated to the right tups so that they are a) in-lamb, and b) in-lamb to the right tup to produce the best possible offspring the following spring. We call it 'lowsing the tups'. At its simplest, it is just about putting a ram, or rams, with a bunch of ewes. Sex, then, five months later, lambs.

Simple. But if you try to breed high-quality breeding stock or tups to sell to other people, and at the same time sustain the character and quality of a good flock of sheep, then it's a whole lot more complicated. Sometimes writers mention what we do as having a kind of 'wisdom of the hands'. They mean it respectfully and as a compliment, as if we are artisans, but I don't like the phrase.

I was back from Oxford for one of my flying visits to help with this work, and it occurred to me that this was more intellectually challenging than anything I had done there for weeks. This was about making judgements, of thinking as well as doing. There was an awkward few minutes of adjustment each time I came home. I might have been daydreaming about something I'd studied that week, or trying to tell my dad something amazing I'd learned, and then I would realize I had botched catching a sheep and got a look that said, 'You're back now. Focus on what we are doing. Or bugger off.' And then I'd switch off the other me and in a few minutes it was as if I had never been away. Shepherds are not thick. We are just tuned to a different channel.

The ewes that run past our legs are in peak condition. They have had eight to ten weeks of holiday after their lambs were taken off, and now are healthy, fat and recovered. Ready for winter. The previous year, my father and I argued about this work, but now we have both changed a bit. If we disagree, I will show more respect, and the less I push, the more he's willing to listen. The Swaledale tup he has bought breeds beautifully coloured sheep, but lacks size, so we are selecting ewes that have the power and quality to match him. To avoid inbreeding, ewes related to the older tups are sorted off and put to new mates.

We know these ewes as individuals, their breeding and life stories, what their lambs were like this year and possibly last. Being away for a few weeks doesn't change that because I was there when they were born and when they lambed, and I was home to help clip many of them. Occasionally, one eludes our memories and we have to check its ear tag and Dad's scruffy old notepad, but a moment later Dad shouts out in triumph, 'It's out of that Geoff Marwood ewe I bought. I should have known her, put her to the big Ewbank tup.'

My grandfather used to have an anecdote for every ewe, and used to drive us mad telling us where each one lambed, what its lamb sold for. Today my father and I take turns offering comment and judgement. We will remember today's judgements a year or so from now and remind each other about where we went wrong (or right).

Sometimes great sheep are flukes, the result of accidental mating combinations. But more often than not there was a plan in someone's mind. I had until recently a fine old Herdwick tup that had bred very good sheep for me. He was old – about ten wives would be sufficient or I would have killed him with the effort. One of my best old ewes had bred a son to that old tup that I had sold for a good price the year before, for £1,900. So as we sorted the ewes to the different tups, I looked for her, found her and made sure she was returned to him. The next spring she bred me one of the best sheep I have ever had. The old tup died the following winter of old age, but I still have his special son to pass his blood on in the flock.

Three years ago a good friend, Anthony Hartley, kindly let me take five ewes to a tup that I admired, but which he did not want to sell. I chose five ewes carefully. It was perhaps the best hour's work I have ever done. Two years later I sold the offspring, the first tup for 5,500 guineas, the second for 2,000, and the third for 950. One of the daughters is now my show ewe and has filled my mantelpiece with silver cups. But, like everyone else, we have bred plenty of average and unexceptional sheep and made plenty of matches that didn't work. As ever with a farming life, the little triumphs matter because of the countless failures.

After an hour or two, we have sorted the ewes into different flocks for each tup. We then introduce the tups. They are strutting and head-butting the gates and stamping their front feet. They know what is happening. Their hormones have been rising within them for the past few weeks. Their manes bristle. They haven't done a thing except eat, drink, sleep, fight and enjoy life in the intakes or valley bottom fields since they were retired from service last Christmas. They will butt each other angrily, or jump on each other. Each ram is raddled (we say 'rudded') with an oily, brightly coloured paste on his chest that leaves a bright smudged mark on the rear end of each ewe he mounts.

A ewe has a cycle of sixteen to seventeen days. We will change the colour on the rams after that period to show us whether the served ewes became pregnant, and are marked again, revealing that they have come back into season. We monitor this so that we can see whether a ram has a fertility problem, and we can tell in the spring which cycle the ewes were made pregnant in. The tup often finds a ewe in season straightaway. He humps her doggedly, like it has been a long time. I make a mental note, because five months from now she should be the first to lamb.

We run the sheep back along the roads to different fields. The ewes gallop away fast, with the tup in pursuit; occasionally he jumps on a ewe as she runs, if he can't wait. In the field, the ewes put their heads down and graze, and the tup starts to inspect each of them in turn, smelling their tails, occasionally kicking one to see if she stands and is in season. He raises his head in the air, smelling the pheromones like a stag. Every day we check these batches of ewes and tups; every other day we catch the tups and refresh the raddle on their chests. If we have any doubts about the fertility or motivation of a tup, we will change him for another. It is crucial that we produce lambs the next spring.

We leave the ewes with the rams for about six weeks (three cycles of raddle – red, blue and green). When they are picked up, they are exhausted, some dangerously so, and in need of R&R. The young tups might mate fifteen or twenty ewes. The best older tups can serve a hundred or more. The field might be twenty acres, and they might circle it several times a day for six weeks in the process of getting that many ewes pregnant, barely having enough time to eat. So, they become exhausted and can lose a massive amount of their body weight. They don't fight so much when we pick them up, because they've used up their energy, and they kind of know that the tupping season is over for another year.

This autumn, as we worked in the fold sorting the ewes to the tups, my best ewe stood up like a statue in front of us, as if to remind us that she is the boss. The best sheep have a sense of their specialness, and this ewe seems to know that she is one of the stars. Her son was the best I have ever sold – maybe the best I ever will (the one that holds the current record price for a shearling Herdwick tup).

I knew he would make a high price when a highly respected shepherd from Borrowdale called Stanley Jackson spent almost an entire day staring at him. I was at Eskdale Show, the 'Herdwick Royal', and whilst everyone else, me included, went about our washing and preening and then showing through the day, Stanley parked himself at the edge of my pen, leaning over the hurdles. Occasionally he would hold his head on one side and scrutinize the tup from another angle. When I asked him what he was doing, he replied, 'Looking for faults.' I smiled and asked him if he'd found any, and he replied, 'Not yet, but I haven't finished looking.' When I won the show with one of my other sheep (the one he liked didn't even get a third or fourth prize rosette), he dismissed that with a wave of his hand, as if it were a trifling detail. Stanley had come to the same conclusion as I had at home the winter before. This was a really good tup, the kind you dream of but rarely achieve.

There is an art to presenting sheep for the autumn shows and sales, and part of that art is accepting that the sheep will not look their best all of the time. Wise shepherds hide their sheep away from critical eyes until they are ready to be seen. My best sheep stay under wraps until the Patterdale and Eskdale Shows, two traditional gatherings at which shepherds have long competed for the pride of having the best sheep in our area, and the whole Lake District, respectively.

One of the valley bottom fields is temporarily turned into a show field. By late morning the pens are full, two long hurdled rows of show sheep. Small marquees flap in the breeze. Other activities like the sheepdog trial and crook competition are judged away across the field. But the important business is trying to make your mark in the sheep pens. The day ends with a lot of banter in the beer tent. The winner is forced to buy everyone else beer from their prize winnings. The prize money is tiny, but no one cares. For many years, Anthony Hartley has taken all the beating in the Herdwick shows. But some of us are working on making it tougher work for him. This year I had the Champion Ewe and Reserve Champion overall, beating a beautiful ewe of his.

The shows build excitement for the sales that follow. The autumn tup sales are the Mecca of our world. For Herdwicks, the two main sales are at Broughton-in-Furness and Cockermouth. For the Swaledale breed, the two main sales are at Kirkby Stephen and Hawes. The whole place tingles with excitement and anticipation. Although the heartland of both breeds is in the north, there are breeders from across the UK and further afield, so everyone descends on these little auction marts in droves.

We enter a number of rams for sale in the preceding weeks and are allocated a suitable number of pens to hold them. The order of sale is balloted so that in theory everyone has the same chance of a prime sale position (being sold too early can be a disadvantage because the 'trade may be cautious'; being too late might work against you if the buyers have secured what they need). Being balloted a pen amongst some of the best breeders can help because of the interest and buzz their sheep generate. I was balloted almost last in, potentially a disaster unless you have a sheep that the top shepherds want, and then it becomes an advantage because they wait all day and are in a frenzied state by the time you are up there, as the possible alternative purchases have all been and gone.

The day started with the pre-sale judging. The tups walked, tugged and jostled to the judging yards, held in a line by their breeders. Everyone is trying to get their tup to stand correctly, as broad and thick-set as they can be made to look, with their heads up. The desired effect is one of arrogance, an alpha male impression, like Russell Crowe in *Gladiator*.

Hundreds of farmers look on, crowding the pens five or six deep. To see anything, they have to climb up the pens in the background. Everyone has an opinion about which tup is the best, and whether the judges know what they are doing or not. And opinions matter because they may become bids later around the ring. After maybe half an hour, the half dozen best tups are pulled into another little yard, and the others dismissed. My tups are dismissed without a prize. I'm not surprised because the two judges, good men, have different taste from me. As I lead my best one away, a friend says, 'Don't worry, that 'un will make more than any of them in the prizes.'

Eventually, the chosen ones are lined up in order of preference, and the rosettes awarded. To win one of the pre-sale shows is, for most people, a once-in-a-lifetime thing; for others but a daydream, though some of the shepherds of the great flocks win with tups regularly.

The rest of the day (or three days for the big Swaledale tup sales) is about the sale. Hundreds of farmers comb through hundreds of pens, trying to find the tups with the attributes they admire or need. The alleys between the pens are crammed with people jostling past each other as they work their way through. Everyone clutches the sale catalogue and works out the breeding and sale order of their favourites. A breeder next to me stands lonely for most of the day, generating little interest, his flock out of fashion. This is a 'tough school' with little sentiment. A cursory look over the pens may be all most people give his sheep.

Other flocks have big reputations, based on many years of having bred well for buyers, and their pens are crammed from morning until night. The tups are pawed over. Feet inspected by pulling back the straw. Teeth checked. Wool parted. Bodies prodded. Ears checked for colour. Hair on their heads felt with the fingers for how 'hard' it is. Everyone here is a scholar about sheep and their breeding. The top shepherds have followed the breeding for decades, and will pour over the flock books that detail the pedigrees and registered rams all winter on dark nights.

'What's he got with [who is his father]?'
'What was his mother?' 'Grandmother?'
'Does it go back to that old Gatesgarth
    tup?'

It is down to whims, fancies and fashion to a certain extent. Men and women chat about what they have seen, and which they like best. Some sheep become objects of intense debate.

'It's just too short...'
'No, it's a hell of a tup...'
'I think it's dirty of the wool in its neck...'
'No. It's one of the best I've ever seen. It'll
    make a bloody fortune.'

This is a world of judgements – some good, some right, some wrong, some bad. Only time really proves anyone right or wrong. It turns out my best tup divides opinion like

Marmite. He is too dark in the wool for some shepherds, perfect for others.

I briefly cause a stir and am subjected to questions about my mental health status by buying the champion for 4,600 guineas in the early afternoon. I like him, though not as much as my own that I am selling – but he is fresh blood and has a lot of style. I can feel Helen glowering at me from the high back seats of the sale ring, so I avoid looking up. The sale rumbles on through the afternoon and the ring starts to clear out as the sale reaches its end. It feels as if the excitement is dying down, that it is too late in the day. I sweat it out waiting my turn. And then I approach the ring, and I'm relieved to see some of the top shepherds have been waiting for my tups. There is a buzz in the pen before they go into the sale ring. An old shepherd I admire tells me it's the best tup he has seen for years. Stanley is across the ring looking nervous, I know he wants him. And then he is sold, and it goes by in a blur. He makes the top price, 5,500 guineas, but just as importantly he goes to one of the top flocks, Turner Hall, where he will be looked after and given a chance to breed with some of the best ewes. For weeks after the sales, I miss seeing him each day, as if I once had a Van Gogh on my wall and now it is gone.

Only tups that have been inspected and

approved by the breed societies can be sold. My father sometimes goes inspecting for the Swaledale Sheep Breeders Association and I sometimes do the same for the Herdwicks. The job is to make sure that no bad faults are missed. Each tup needs, obviously, two testicles, good teeth, sound legs and feet, and to be the appropriate colour for the breed. Judgements are made about relatively minor issues because self-respecting breeders rarely put a poor sheep in front of the inspectors.

'It's a shame but his teeth are just slightly over the edge of his pad ... I think we'll have to turn him down...'
'By he's a good tup, but you'd have to say that leg is a little bit twined [twisted]. I don't think we can pass him... Sorry, lad.'

Being one of the inspectors calls for the diplomatic skills of Henry Kissinger. You risk upsetting the breeder if you turn the sheep down, potentially ending your chances of ever selling him sheep again. But pass a sheep with a fault and it is likely to be noticed and brought to other people's attention later at the sales. The whole point is to protect buyers at the sales, who should be able to buy with confidence, knowing that men who know what they are doing have okayed the sheep. So the inspectors sometimes do a funny little set-piece scene in which they spot a fault and look uncomfortable; they have a bit of a steward's inquiry, and usually the breeder, to save their embarrassment, will intervene... 'Don't worry, it's not right, I never realized it was that bad. Turn it down, you'll have to.'

The inspectors, freed from their embarrassment, fail it and move on to the other tups.

If I had only a few days left on earth, I would spend one of them inspecting Herdwick tups. The inspectors are driven around the Lake District valleys to all the beautiful little stone-built farmsteads. Some nestle under rocky crags, and all surrounded by endless miles of walls that track up the fell sides and carve the valley bottom into irregularly shaped meadows. As you approach each farm, the 'coordinator' who lives in the same valley will tell you about the history of the farm and the family so that you understand the people and place before you.

'This was once one of the greatest Herdwick farms ... my father said there wasn't another flock to match it... But the son was no good ... when he left, the National Trust put in some daft bugger from down South and the sheep were wasted... But there are still some good 'uns ... and this new lad is trying to turn the job around ... they say he has a nice one this time.'

Each farm has its own stories that are sustained only in the memories of the other farmers and shepherds. Even individual fields or bits of the commons have names. We know most of these folk, but often you will not have had a reason to visit their farms before. At each place, you are made welcome, though people are nervous as the decisions could spoil their autumn sales. Out of the farmhouse will come the whole family and everyone does their hellos. We are almost always asked if we'd like a cup of tea and a bite of cake. Then we are ushered to the sheep. They will be penned in yards that are little changed from when Beatrix Potter bought some of these places. The ironwork on the gates is often worn shiny with use, and the timber rails smooth and often red from the tups.

Standing before us there might be a dozen (sometimes many more) Herdwick tups. These tups are charged with passing on their masculinity to their sons, so any sign of softness, or standing in a feminine way, is frowned upon. They should stand four square, or as we say, 'a leg in each corner', like a sturdy oak table with chunky legs. Their white heads glisten in the summer sunshine. About half will have curly, powerful leg-bruising horns; others without horns are called 'cowed'. The kind of white in the head and legs matters, and any sign of dullness or grey, or large black spots, is frowned upon. You can often tell the quality of the breeding from the whiteness behind the front leg in the 'armpit'. It is just weeks since they were sheared so their powerful grey bodies are thick and long and athletic.

First thing that goes through my mind (and everyone else's, if they are honest) as I walk in the pen as an inspector is, 'Have they got a really good one that I might get hold of before anyone else does?' There are dozens of little things I am looking for: practical things like the size, healthiness, alertness, mobility, legs, fleece and teeth. Without these, the sheep cannot live on the fells. But because sheep are cultural objects, almost like art, I'm looking for style and character as well, and finer breed points, like how white their ears are. White lugs won't help them survive the winters, but they will help me breed sheep that I can sell to discerning shepherds. The little aesthetic things become the symbols of good breeding over a long period of time.

I know exactly what the perfect Herdwick tup looks like because it struts around in my head. I measure all my real ones against it and know exactly how far short of it they fall, all too often.

Once the work is done, we inspectors and farmers discuss the breeding of the tups and other farming related topics such as whether the hay has been got in. Then you depart off to the next flock. We might inspect a hundred tups in a day, on maybe fifteen farms. It might take ten days to cover the whole breed area, so different inspectors do a day each, generally. Endless debates take place about whether the inspectors are consistent on any given issue. Like football referees, their judgement is scrutinized and not always respected.

I'm on the third or fourth floor of a building just off Oxford Street in London. I left the flat in Oxford at 5.30 a.m. to catch a train, and won't get home until 10 p.m. My work cubicle is about three feet by four feet square. The shelves above the Mac I'm working on tower up towards the ceiling, covered with the assorted papers and other rubbish of the previous occupant. The nearest window is about twenty feet away, but it hardly matters because there isn't anything to see from it except the back of the neighbouring building. There is nothing green to see out there anyway, except a sickly looking little tree in the square below.

I'm working as a sub-editor, despite having zero experience. After a term or two in Oxford, I realized that to get the kind of well-paid job I needed, I had to get some 'work experience'. I secretly fancied myself as the next Ernest Hemingway, so I thought maybe I could be a journalist. I'd applied to some magazines for work experience. Only one replied and I was called down to London to have a chat with the editor.

It didn't start very well, because when I arrived I had to use an intercom system, and not knowing how they worked, I pressed the buzzer continuously whilst I was talking to the person upstairs. They told me angrily that I could stop pressing the buzzer. The whole office buzzed each time the intercom was on and stayed buzzing as long as the button was pressed. When I got up there, everyone peered over their computers and smirked. The editor was very friendly and seemed to realize I was way out of my comfort zone, but kindly agreed to give me a chance. When I came back to start the work experience on the date agreed (some weeks later), I passed a man leaving the building with a cardboard box under his arm, looking flustered. He bustled past me in the doorway. When I got into the office, I was told to sit in the cubicle and wait. I waited about three hours.

Then the editor emerged from her office and thrust a few sheets of A4 with scribbled notes on at me, and said, 'Proof that.'

'But...'

'Sorry, I haven't time to talk ... just do it.'

I had a funny feeling that she didn't recognize me at all. When I took it back to her half an hour later, she was on the phone and simply took the papers from me, motioning intensely for me to keep quiet. She then handed me another piece of paper with scribbles on. I left at the end of the first day feeling completely baffled.

Over the days that followed, I began to pick up a few things, not least that sub-editing has its own language of squiggles that you use. So I learned those, and did the best I could. It was a crazy manic atmosphere of a kind I'd never experienced, often interspersed with hours when I had done the work but could not get anyone to tell me what to do next. I'd be waved away by the editor or another member of the staff. At lunchtimes I'd sit on a bench in the square and marvel at the beautiful girls flooding out of all the fashion magazines and fashion houses in that district.

After a fortnight or so, I was beckoned into the editor's office. The magazine had gone to press and the atmosphere had changed. She asked how much they were paying me. I explained that they weren't. I was there on work experience, and no one had offered to pay me. She seemed surprised. Then she explained that I was doing the job of the sub-editor who'd been sacked a fortnight ago, the man leaving with the cardboard box on the day I had arrived.

She asked me to stay the whole holiday and come back the next summer. The following summer was the only one I have ever spent away from our farm. It was the weirdest few weeks of my life.

I didn't know anyone in London, and I never wanted to be there. This was not how my life was meant to be, but needs must. It was as if the gods were showing me how tough everyone else's lives were, and what I had left behind. I understood for the first time why people wanted to escape to places like the Lake District. I understood then what National Parks were for, so that people whose lives are always like this can escape and feel the wind in their hair and the sun on their faces.

I promised myself that I would be at home the following summer. I was, but not under the circumstances I'd imagined, because in 2001 the Foot and Mouth epidemic broke out.

From the high ground where we feed our ewes and lambs, for as far as I could see, there were towers of smoke rising from pyres of burning sheep, cattle and pigs. The land was shrouded in a grey haze. The wind carried the sickly smell of burnt flesh and the chemical smell of the fires. For weeks we were under siege. Those of us who had not yet been struck down by the virus were waiting to be hit by it. Our landscape was riddled with it now, because the government was too slow to react to its spread in the beginning, completely oblivious to a farming world where livestock moves around the countryside (as it always has). The TV news showed a map of the cases spreading, an ugly grey stain that seemed to cover my whole universe. The solution decided upon was to clear certain zones of livestock so the disease could be contained. The land would be cleared of sheep initially, but the cattle would be left in their winter housing.

They came to collect our sheep at lambing time. We loaded pregnant ewes into the wagons. The few lambs that had been born were loaded as well. I have never done anything that felt so wrong, so against everything I was ever taught to do.

The auctioneer sent to value them for compensation cried and said it was 'criminal to kill such good breeding sheep'. Many of the sheep were descendants of the good ewes my grandfather had bought in the 1940s. Sixty years' work wiped away in two hours.

Our cattle later got the disease anyway and were shot in the fields by a police sniper: killed one at a time, with a crack of a rifle, until the fields around the village looked like something out of a war movie. The villagers stood on the green, watching in disbelief. My neighbour stood with a shotgun at his field boundary, ready to shoot any cattle that threatened to jump his fence and infect his 'clean' cattle. He apologized, but said he had to protect his stock. I told him I understood.

I'd have done the same. My dad wanted nothing to do with the whole miserable business and went in the house, leaving me to supervise amidst the chaos. I felt dirty and ashamed. At one point I turned to someone in disbelief and said, 'Is this really happening?' They replied, 'I think so.'

After it was over, the slaughter finished and the men gone, I walked around the farm in the evening sunshine in disbelief. It was a beautiful evening in the English countryside, with a peach-red sunset, but the fields were speckled with our dead cattle. Red cattle. White cattle. Black cattle. Strangely peaceful, they lay in all sorts of mangled and contorted ways. I knew those cattle, so it was like seeing old friends dead. The setting sun created all kinds of grotesque shadows. My mind couldn't quite process it. It was surreal, like I was watching a movie. The farm was eerily silent, something we had never known before. The next day our dead and bloated livestock were loaded by diggers like trash into wagons and taken to a hole in the ground miles away. There was a look on my father's face of pure disgust at this whole spectacle.

When the last wagon had gone, I went into the barn, away from everyone, sat down in the shadows, held my head in my hands and sobbed.

Then the farms were empty. And we didn't

know what to do without livestock to care for. I waited to hear my dad getting up, but there was nothing to get up for. Our sheep and cattle were dead. Someone had pressed 'Pause' on our way of life, and we weren't sure if anyone would press 'Play' ever again.

Our farm in the fells was one of the last farms towards the mountains culled during the epidemic. Had it spread west by a few more fields, the disease would have got on to the unfenced Lakeland fells where it would have decimated the ancient hefted fell flocks on the commons. Ninety-five per cent of the Herdwick sheep in the world exist within twenty miles of Coniston, and were at grave risk of being wiped out. But an essentially urban government didn't understand. To them, a sheep was a sheep, a farm simply a farm. The idea that something precious was on the edge of destruction was never really grasped.

We are often in the hands of other people, our fate in the hands of shoppers and super-markets and bureaucrats. In the end, the fell flocks mostly escaped (though many young fell sheep were culled when they were caught in the killing zone on their wintering grounds in the lowlands). Many great flocks of sheep and herds of cattle were destroyed, but thankfully not all.

I stayed home all that summer. We hired a bunch of my mates and cousins to undertake the massive job of pressure-washing the entire farm until it was spotlessly clean to the government inspector's satisfaction. Without livestock on the farms, everyone and everything was different. People you knew who had never relaxed in their whole lives suddenly were thrown out of their old ways. And the farms were clean and that was disconcerting and made them feel strangely clinical and dead.

Pubs and restaurants suffered as well because people assumed that everywhere was closed, so the visitors didn't come that summer. There were tensions, too, because farmers had got different auctioneers to value their stock for compensation purposes, and some had valued higher than others. People felt cheated. Perhaps the worst off financially were those farmers that hadn't got the disease, and so were not compensated. Unable to sell their livestock, their businesses were effectively frozen for months, with costs mounting and nil income.

But it was not all bad. There was a certain community spirit that thrived in those circumstances. Our farm had probably not had so many people working on it for decades, and once we got past the grimness of what had happened, we had quite a lot of fun working together. Soccer matches after work. Nights out at the pub.

In the months that followed, my mother and father quite sensibly gave notice to leave the rented farm (because it was losing money some years after the rent had been paid), and left a year later. They bought a house on the edge of the local town and farmed my grandfather's land remotely. We would keep the farm in the fells and see what happened. I worked with my dad for months during this time.

Leaving the farm is supposed to push you into having another life, but my leaving had just made me realize that the farm was the beginning and end of everything for me. When I was young, my grandfather had stood with me in a barn that was isolated up in some of his fields, and said that someday I should make it into a house and live there. And now that idea was in my head. It was my goal. It was the first thought in my head in the morning and my last thought at night. As the joke goes, it isn't a matter of life and death, it is far more important than that.

Because all of the farms later had to restock at the same time, prices became inflated. We were cautious. We were not sure what to restock our farm with, so we bought some Herdwick draft ewes from Jean Wilson. They transformed from old, tired-looking creatures when they got on to our grass and thrilled us with how well they had done. One day, as we were working amongst them in the sheep pens, Jean came to talk to us. She told us that some of the ewes were good enough to breed 'pure' (instead of crossing them to produce lambs just for meat). So she brought us a distinguished old Herdwick tup to use. Jean is formidable, so we did as we were told. The next spring we had our first Herdwick lambs. Those old ladies were the start of our current flock. Two of them turned out to be very fine breeding ewes. One of their first lambs grew up to win our local sheep show and gave me the bug for breeding them. One day I said to my dad, 'I think I'll breed the Herdwicks.' He smiled and said that was fine. Since that time we have specialized in different breeds. Today, I have the granddaughters and great-granddaughters of those old ewes in my flock. My current farming life was reborn during those sad months when it seemed as if everything was broken.

Winter

About living in the country?
I yawn; that step, for instance–
No need to look up – Evans
On his way to the fields, where he hoes
Up one row of mangolds and down
The next one. You needn't wonder
What goes on in his mind, there is nothing
Going on there; the unemployment
Of the lobes is established. His small dole
Is kindness of the passers-by
Who mister him, who read an answer
To problems in the way his speech
Comes haltingly, and his eyes reflect
Stillness. I would say to them
About living in the country, peace
Can deafen one, beauty surprise
No longer. There is only the thud
Of the slow foot up the long lane
At morning and back at night.

R. S. Thomas, 'The Country',
in *Young and Old* (1972)

He sees me before I see him – a young, bull-
necked raven. Coal black. Scared of nothing
and with a bellyful of the dead. Ravens live

on our failures. Brutal. Arrogant. Cruel. And sometimes stunningly beautiful.

I am reading an ear tag for my records, and write it down in a soggy and dirty notebook, '15547. Dead. Pneumonia.'

Had I turned the corner with a shotgun, the raven would have risen over the wall and skulked off to a tree, just out of range, with a laughing and throaty *'Kraark'*, but he is knowingly disinterested in a man armed only with a ballpoint pen. His thick, black, hoary neck ruffles as the wind catches his feathers. Greedy. Delirious. He rises like he has a stone in his belly, punch-drunk on carrion.

Our casualties are not pretty, because life and death aren't. Winter is attritional on the flock. Two aged ewes, too old for this winter, bellies bloated and eyes stolen, lie in the yard. They lie next to a young vixen with a hole blown through her belly, insides almost outside, a resentful fang bared on her gnarled wild face.

Atop the corrugated roof of the bullock shed, the raven steps from one claw to the other. Every movement of his thick dark body says he is gorged. On laboured wings he departs into the darkness.

There are moments like this, when you are half-beat and dark news shadows menacingly over you.

One of these ewes lying dead means a lot to me. She is the best I have. She is like the matriarch of the flock. She led the flock out of the snowdrifts last winter when they were in danger.

Snow. Shepherds fear and loathe deep snow and drifting winds. Snow kills. It buries sheep. It buries the grass and makes them even more dependent on us for survival. So we suffer everyone else's excitement. Snowballs. Snowmen. Sledging. We fear it. A little snow is harmless, we can hay the sheep and they can endure the cold easily enough. But the combination of wind and deep snow is a killer. It kills sheep, and can easily kill men and women. If you've ever seen ewes lying dead behind walls after the snow has cleared, or seen lambs lying dead where they were born, you will never love snow so innocently again. Still, as much as I fear and loathe its worst effects, it does make the valley beautiful. White. Silent. Cruel. It muffles all the usual noises, save only the wind-like cry of the beck, a little quieter than usual. I can tell it is deep snow before I open my eyes just by the missing noise. But a clock is ticking in my head, telling me that until I have seen and fed all the sheep, my work will not be done.

I step out into that Brueghel painting of the snow and the crows. Oak trees and thorn dykes stand out in the white like black coral. I feel alive, necessary, needed. I have to be my best self today, fight what has happened, or creatures will go hungry. The snow is heavy now and layering up fast on the land. Leading the hay to the ewes on the quad bike in heavy snow, I become white myself. Thick white snowflakes carpet me as I head up the road. You see them falling by the millions like duck-down feathers. Some land on my face, crumple into my warm eye sockets, and blind me with a soft wetness. I feel the lightness of a snowflake land on my tongue. Soft. Fat. Delicate. Like the Snow god had placed it on my tongue for Holy Communion. The quad bike tyres make a crunching sound, packing down the snow on the road.

The field gate I open has a three-inch layer of soft snow across its top bar. The first flock I go to feed are away in a ghyll where their mothers and grandmothers have taught them to shelter when the gales come in. Mountain sheep have a sixth sense for the weather on their own territory. I find them under Scots pine trees, forty feet beneath the danger of wind and drifts. The oldest ewes will have led them here, and will stand stubbornly if the younger ewes try to lead them out to danger. The flock takes its cue from the elders. They know they are safe here, with tussock grass to chew on to keep them alive if the snow lasts for days. This place is almost as good as a barn, windless and watered by the beck that still carves the ghyll out of the mountainside.

I throw emergency rations of hay down the sides and they gather to it. The ewes tug a gobful away from the slices and start to chew. With every mouthful I see them eat, I loosen up. Ewes that are sheltered and fed with some dry hay can survive here for days and days. I count them and learn there are two missing. But then, suddenly, they too are tumbling down for the hay. Relief. These two young ewes have been to scratch through the snow for sweeter grass. They will be OK now. They will hold to the hay through the snowfall.

But there is no time to dwell here admiring the scenery. I have other flocks to feed. The snow still falls heavily, and the valley is changing about me.

Whiteout. The road in the distance is silent now. Empty. The valley is being cut off from the world. I hear my father shouting to sheep on the lower ground where he is working. Snow ploughs will be working soon, but it might be a week before they get here. They will focus on the motorways and towns. I am already fretting about the flock of ewes furthest away on some high ground, I'm not sure I will get to them if the snow keeps deepening this fast (and getting there is just half of the problem).

I take them some hay so they can endure

the snow with something good in their bellies. I need to get there quick. The quad bike labours, skidding and sliding, occasionally lurching sideways. I drive through the village, past cars being pushed into drives by folk who've just returned from trying to get to work in the local town, beaten by the snow. I wind up a little lane that leads to the higher ground. But the snow is packed down like ice and I can't get up the hill. I turn around, determined on another way to get there across a field or two. I pass my neighbour doing similar work. A little nod says he's seen me and knows where I'm going. That little nod might keep me alive later. No one else knows where I'm going.

The snow is getting deep now and I have to concentrate or I could hit things hidden beneath it. Troughs. Branches. Stones. Soon I am at the field where my flock should be, but I cannot see them. They must be sheltering behind a wall across the field, but the gateway is drifting and I can't get the bike through it. I have to find them. The distance is small, but trudging through the snow with a heavy load makes it feel epic. Floss leaps through the deep snow beside me as if she is jumping waves. She knows what we are doing, and gets to the wall before me. She runs up the drifting snow against the wall to see what's over the other side. She looks back, impatient for me to catch up. We find some of the ewes quickly. Coated in snow. Faces white. Their black friendly eyes seem pleased to see me, their wool insulating the snow that lands on them from the heat of their bodies. They rush to my legs and start on the hay. I count them, but it is hard because other ewes are emerging out of the blizzard from all directions. I struggle to get a decent count, but some are missing, maybe a dozen. I have a decision to make... If I stay here much longer, the quad bike will get stuck in the lane and I might get into all sorts of trouble and might not get back for the other flocks. Then they appear from out of the whiteness.

I don't like this snow. It is layering up into drifts very fast. I decide to take the sheep away with me, get them lower down to some shelter. I need to hurry. I push the ewes through the snow, but they want to go back. So I pull an empty hessian feed bag out of my pocket and try to persuade them to follow me. If we can get a few hundred yards down this hillside into a new field, there is shelter. I fall on my backside, and get up again. I trudge my way through the growing drifts and am pleased to see that the ewes seem to understand.

The best ewe follows me in the trodden path I am making. She has bred me great sons and daughters, helping to make my flock better. She has a sense of her own importance at all times. She was shown as a young ewe and for years afterwards would show off when I took people to the fields to see the flock, standing like a statue. In summer, she leads them down the lanes from the high ground and across the beck, jumping right over it, with the whole flock following through the air. She is canny and streetwise; she knows they are being led away from danger now.

I send Floss back around the rest and they follow in Indian-file. I am sweating, but freezing cold in my toes and fingers. I will go back for the quad bike later and fetch it down the wind-swept fields in another direction. We reach a gateway that is deep with snow, up to my waist and getting worse as the wind drives more snow across the land into any slack places. Beyond this point the ewes will be safer, but I can't leave them here in this lane, so I crunch through almost up to my chest. I wonder if this is a good idea, but the old ewe is already following in my footsteps. The others look at her, unsure whether to follow. But then one of her daughters comes, and they all bunch up at the beginning of the little white gully I have created. And then I am through the drift, the ground reaching back up towards my feet. I tumble over as I hit a stone, and the old ewe walks over my legs, followed by eighty others, all of which are now on a mission. They trek away down the fields to where the snow is less deep, and where I go and feed them with hay. Whatever happens now, the sheep can endure it. They are safe here, out of the drifting winds. Floss comes and licks my face, she knows I need her today in this blizzard.

I eventually get the quad bike out of the snowy fields, and go home. My hands are numb with cold now. I'm heading to the warm tap. The house door has a little drift against it and it falls into the kitchen as I enter. The children are excited to be off school and want to go sledging and beg me to take them. I groan.

Helen curses me for making a mess. I tell her all about what has happened, and she teases me about how much I love that old ewe. She calls her the 'Queen of the Flock'. And then she sees how cold I am and fusses.

Winter is my swollen pig-like fingers throbbing under the hot tap, thawing out, as I howl unheard blasphemies at the stinging pain. It is my bloodshot eyes in the mirror as I finger out hayseeds. It is snowflakes or hailstones hitting my face as I drive the quad bike into the wind, snow or rain becoming perfect warp speed lines like those scenes in *Star Wars* when they flick the throttle and the stars transcend. Winter is my father's neck in front of me, streaming with rain as we catch an old ewe that is unwell. Ewes grabbing desperately at hay in a storm before the wind robs them of their rations. Lambs lying dead, defeated before they have even started. Winter is hayracks and trees blown over, torn and smashed.

Winter is a bitch.

But winter is also pure brilliant cloudless days when all is well in the world – when the fields dry out, the sheep are at peace, full of hay, lying in the sunshine, and we can work and also enjoy the beauty of the valley and its wildlife. Winter is beautiful too.

Little things you see make it special. Skeins of geese pass over, high in the frosty blue. Ravens tumble over each other down the wind, like a black ribbon descending from the fell. Foxes skulk across the frosted fields at first light. Hares watch you with big dark watery eyes.

The next day, I return and find the ewes. They've been buried behind a wall and are tasselled in handfuls of snow that weigh them down. But they are OK. The drifts here are smaller than higher up the mountainsides. I throw them hay and count them to know all are safe.

We didn't lose any sheep in the blizzard, but in the weeks that followed we lost more than we would normally. They were worn down, and they paid the price weeks later at lambing time. Some farmers we knew had hundreds of sheep buried for days and days, and lost dozens. Our neighbour spent a week clearing the road to get to his ewes with a tractor and loader. In Wales, Ireland and the Isle of Man it was even worse.

A week or two later, eighteen red deer corpses were found in a frozen tangled heap in the next valley from us. They had descended from the fells to escape the worst of the blizzard, sheltering below Grey Crag. The drifting snow had curled over the top of the wall that they had sheltered behind, entombing them. The ground beneath their feet was grazed bare and deep in their shit. They died hungry, cold and dehydrated. The melting snow revealed them to a shepherd friend of ours.

I am looking in through the glass at a carving from a reindeer antler from at least 13,000 years ago of some swimming reindeer in the British Museum. I am spellbound. It reminds me of the animals that the shepherds here carve into their crook handles. The carving had been found when a railway was built past a rocky cliff in the Pyrenees in the 1860s. It throws me, because it makes it clear that the 'North' has always been moving. It was once hundreds of miles to the south. When this was carved, our landscape was still under the ice. That beautiful little carving from a reindeer antler gives me a glimpse of the people who travelled north in the summers to graze across landscapes like ours, a people with a sense of grace and beauty, who probably stopped and looked, and saw things they found beautiful. These were a people that imbued animals with great meaning.

As the ice retreated, nomadic hunter-gatherers crossed our tundra-like land following herds of wild animals. If you could have looked down on North-West Europe from something like the International Space Station, say 16,000 years ago, and fast-forwarded the history of Earth below, you would have seen the ice retreat, a white tide going in and out, slowly backing northwards from places it had stubbornly subdued, while gradually losing ground over the generations – the 'North' retreating slowly back towards the North Pole. You would also have seen that, with so much of the oceans trapped in vast ice sheets, sea levels were much lower than today. This little corner of North-West Europe was just a tiny part of a bigger land-mass.

The landscape shifted as you moved southwards from deep glacial ice, through tundra to steppe and then forests, furthest from the ice. As the ice retreated, a new kind of land emerged, and things started to grow that couldn't have survived in the ice and snow. Trees marched north gradually, imperceptibly, until this became a landscape of trees up to at least 2,000 feet on the mountains. And behind the retreating ice came the animals of the tundra, and later of the forest – reindeer herds, wolves and bears. For several thousand years after the ice, the small number of people here weren't farmers, but hunter-gatherers. Then there was a period (four or five thousand years ago) when they were partly settled farmers and still partly hunter-gatherers. Then, finally, by about three thousand years ago, they look to me as if they had become settled farmers, people that I could feel an affinity with, even if their lives were different from mine. There were successive waves of invaders in the centuries that followed, but none that broke it. By a thousand years ago, the farming looks very familiar indeed. After that, the scale changes, but the structure is essentially the same. The landscape now is the same landscape Wordsworth wandered through.

No one can be sure, but there is a suspicion that the fell people just carried on beneath the waves of 'history' that fill the story of England. Sometimes it feels to me as if the tide of the North receded with the melting ice and left us in place in the hills, little islands sticking up from an encroaching sea of southern 'civilization'.

I could tell from three hundred yards away,

as I approached the field, that the old ewe was ill. She looked different. It is weeks since the snowdrifts melted, and she has been in good health, but now she looks hollow and has an ear down. And although I did everything to save her life, she deteriorated quite quickly and died a few days later of pneumonia. It wasn't the snow that killed her, but the wet weather that followed.

We are not sentimental people, but we share our lives with these sheep. We care about them. That ewe was born on our farm seven years ago. Since I bought the Herdwick ewes, I have built up a flock that lives on our intakes. Although Herdwick sheep are predominantly fell sheep, there are a number of flocks beneath the high fells. Some of these specialize in producing tups to sell, taking advantage of their enclosed land (which lets you control the breeding more tightly than you can on a high fell farm, where lots of tups have to be released, together with many hundreds of ewes) to have a smaller but high-quality flock. For the past ten years, we have kept some of our best male lambs from those Herdwick ewes and sold them each autumn to other farmers. I have been working up through the ranks from an amateur, making lots of mistakes, to being taken a little more seriously now as my flock has improved.

The old ewe was a big part of that journey because she was my best. I remember her being born, because her mother was my show ewe as well. She was born under a fallen tree out of the wind and rain. A single lamb. She lived on our hardest ground all of her first summer, a place she was always happiest to return to. The first autumn she was chosen to be one of the best females of her age, and kept when some of her peers went to the sales because they were surplus. She went away the first winter to the lowlands to graze on a dairy farm with lots of grass, and grew into a fine young sheep. The next spring she returned to her 'heaf'. She won her class at our local show, and so did her first tup lamb. He was sold the next year for £2,000 to Joe Weir at Chapel Farm, Borrowdale. There are sheep that are descendants of hers in the other valleys now. Her last daughter to be born has taken after her. She, too, is a show-off, and likes to lead the flock wherever it goes when we are moving them. She also stands like a statue when anyone is watching. I dream that she will breed well, that the family will go on in my flock. These little dreams sustain this way of life.

Life and death are part of the work on a farm. It used to be that all farms had a 'dead heap' or a 'dead hole' where the bodies were thrown. We were supposed to dispose of the carcasses by calling a 'knackerman'. He'd turn up, fag in mouth, on an old wagon, trailing the smell of death through the country-side. I used to wonder who in their right mind would choose that job. But someone had to do it.

One day, we took a dead ewe to the knacker's yard. We were in my dad's battered old Land-Rover. Blondie's 'Atomic' was on the radio. I was used to dead things, but I'd never seen anything like this. There were piles of bloated cows and sheep, tongues thrust out, eyes poking out of their heads. Fat black flies were everywhere, puddles of drying blood and bile, pools of piss. The smell made you retch and was so bad it followed you home. It was like some vast panorama of animal death by Damien Hirst.

A man was sitting on the bloated belly of a large black and white cow, his bait-box resting on the cow's belly under a cloud of flies. Huge bluebottles were crawling on his hands, which were covered in dried blood. He was eating a sandwich of white bread, butter and thickly sliced boiled ham. And he had a comical grin on his face.

We chucked the ewe out next to a small mountain of other corpses, our boots rimmed in a grey slime that they sank into. As we left, my dad, usually unflappable, said, 'Christ, did you see that daft bugger's hands ... and eating a sandwich.'

When I finally returned home from Oxford,

my family and friends were proud that I had 'done well'. But when they said that, I couldn't help thinking that I hadn't yet done anything at all. I was unemployed; I had a student loan to pay back; and there was no house on our farm for Helen and I to live in. I should have been worried, but I wasn't. I was elated.

As the Lake District fells rose up in front of us, I felt that I was home. I could feel those fells encircling me like friends, and I punched my fist and shouted 'I AM HOME'.

Helen laughed at me and said I was crazy.

I'd gone to prove a point to myself and maybe to other people too. But there wasn't much satisfaction in it. I'd lost the hunger to keep proving it.

Grey clouds pass overhead. I am picking stones in the middle of a vast brown field. My job is to drive a digger, stop every thirty or so yards, and throw into the loader bucket on its front any stones that the plough has turned up. I am working on my cousin's farm. He drives past and teases me that he's never had such a well-educated slave. I laugh and tell him to fuck off. I'm grateful for the work.

Within a day or two of getting back from Oxford I was soon getting plenty of offers. Walling. Clipping sheep. Milking cows. Picking stones. But none of this work pays enough to buy a house here, or to get a mortgage to convert the barn on my grandfather's farm. I knew I needed a 'good' white-collar professional job. I could work nine to five, keeping the holidays and every weekend for the farm, plus a couple of hours every morning, some lunchtimes as I was passing by, and every night. I could do a lot of farming every week, and be there most days. It meant switching between suit and farm clothes every day, but I hoped that in about ten years we would be able to do the things we wanted to do: build a farmhouse, and keep the farm going.

Since Oxford I have worked on the farm with my father, but I managed to fit that around another professional life to earn a living I got a series of jobs focused on the economics of historic places, and realized I was fascinated by the subject. The internet and smart phones meant I could work from home a lot and with flexible hours. Today, I am an 'expert adviser' to the UNESCO World Heritage Centre in Paris and work with them on a freelance basis to help ensure that tourism benefits host communities. One of my shepherd friends says I'm 'a bit like James Bond', I go to lots of random places and no one knows quite what I am up to.

Sometimes I am doing my other work while I am standing in the sheep shed. Access to the internet and smart phones means you can literally be anywhere (even surrounded by sheep) and no one else needs to know. A colleague on the phone might say they just heard a sheep. I tell them they imagined it.

My other working life has allowed me to build a farmhouse on our farm.

Twenty years have passed since my grandfather died and the farm was in trouble.

After Oxford, Helen and I ended up living in Carlisle for a couple of years, thirty miles north of the farm. I'd leave Helen each morning to go to the farm, or to work elsewhere to earn our living. When I got back, she would thrust our new baby into my hands and say 'your turn'. Next door to us lived a lovely old couple. The husband, Fargie, called water 'Council pop', because when they were young they couldn't afford anything else to drink, and his mother had made a joke of it.

Later, we moved to a village not far from where I grew up in the Eden Valley. My friends teased me that we were moving very slowly back towards the farm but that it would take about three lifetimes at the current rate of advance.

Helen loved the house we lived in, in the village of Newby. Our second daughter Bea was born in the bathroom there, and the next day an old neighbour of ours came round and told us she was the first child born in the village (rather than the local hospital) since he had been born seventy or so years earlier. Helen didn't really want to leave the home she made there. She was worried that we were moving to a farm in the middle of nowhere, away from her friends and neighbours and a life she had built, to an old barn in the middle of a field, to start again. But moving to the farm was always my dream. Helen accepted it because she loved me, and was loyal to things I needed to do. She was farming born and bred, but, like many sensible daughters of farmers, she'd kept a step removed from farming. She now jokes that it took me nineteen years to persuade her to show any interest in farming, but now she does a lot of jobs on the farm and knows more than she lets on.

In the end, we were able to convert a barn on the farm into a house, and this is now our home, and our children have settled into the local school. My whole world is now on that farm. My family. My sheep. My home. Even on the endless wet, grey days, I never regret being here – which is good because we get a lot of them.

Sometimes it can feel like Groundhog Day, especially in winter. After the autumn sales, there is an ominous morning-after-the-night-before feeling in the air, with winter stretching out in front of us. Cold and wet weather can start as early as October and last right through until almost May before it feels warmer again – a whole eight months of the year that tend to feel like winter. We rarely get the seasons as they do in the south of England. Spring and autumn are often hurried transitions, never really equivalent in length or spirit to winter. Only summer sees the world relax a little.

I wake to the sound of wind and rain scolding at the window. I can see from my bed a dirty brown rug of heather, mud and skeletal oak trees. The beck roars away in the ghyll tumbling over the stones. The fells stand capped in a dirty smudge of cloud. That tiny moment of looking out tells me what my day will be like: whether I am to have an easier time in walking boots, or whether I will be fighting through in layers of warm and waterproof clothing.

From the moment my eyes unfasten, a clock is ticking in my head, marking time, telling me daylight is limited, that the flock have not been so lucky as me and have endured whatever the weather has thrown at them through the night. Powered by guilt or shame, the voice is telling me not to screw up. Daylight is limited in the winter months. As the sun crests above the fells to the east, I know the timer has started ticking. I have a finite period of time to get around my stock and squeeze in whatever other jobs need to be done. On good days I don't notice it. On bad days the clock ticks heavily in my skull. There is no opt-out. Something might die because you couldn't be bothered.

There is little joy in working in the sodden wet days of winter. It proves too much for some people. When the National Trust rents out farms to people new to this world – folk that are often giddy with enthusiasm for the farming life – it tends to end badly. The get-up-and-get-out voice in their heads isn't strong enough and they just don't care enough about the sheep and the land to sustain their initial enthusiasm once the going inevitably gets tough. Things fall apart, and they soon leave. The voice in our heads is what holds the Lake District together, puts the walls back up, drains the fields and keeps the sheep well-tended and bred. Many of these things defy rational economics. Some of our friends spend maybe fifty or more days a year rebuilding the walls on their farms, when letting them fall down and selling the stone might be the modern solution. It is done because it should be done.

I grab breakfast – Corn Flakes or porridge.

There is no such thing as bad weather, only bad clothes. That's what they say. I'm not totally convinced, but anyway, I layer up. Thermal underpants. Vests. T-shirts. Until I look like a pass-the-parcel version of myself.

Chunky. Wrapped. Warm. I have that sinking feeling that now, at 6 a.m., this is the warmest and driest I will be all day. Keeping dry is the biggest challenge, so our farmhouse kitchen is a mass of sodden jackets, overalls and hats and gloves. The room has a slightly humid, damp and sheep-laden smell. The clothes are never quite dry. I've had pneumonia and no one will be surprised if I get it again. It would once have killed men and women here in their damp little houses. I wear out good jackets faster than we care to buy them, they are so quickly torn, ripped and shredded. I look like an old man when I'm farming. I look like farmers here in ancient black and white photos.

My job is simple: get around the fields and feed and shepherd the different flocks of ewes – dealing with any issues that arise.

First rule of shepherding: it's not about you, it's about the sheep and the land.

Second rule: sometimes you can't win.

Third rule: shut up, and go and do the work.

There is a moment each December when the ewes start to need supplementary feeding with hay. The sheep lose condition as the harshest weather slowly takes its toll on the flock. We endeavour to reduce the negative effects by feeding and tending them, but there are many days when I know that, despite my best efforts, the sheep are in a worse condition at the end of the day than at the start, rain-lashed, covered in snow, up to their knees in mud, or sheltering sullenly behind the walls from the lashing wind.

As a boy, I would go out with my grandfather and help build the hayracks that he constructed from fence posts and wire netting. I would hold the fence posts whilst he knocked them into the crusty chilled ground. As the hammer fell, you would wonder each time how good his aim was, and soon learned to hold in such a way that you could withdraw your arm quickly if he missed or slipped. He would laugh and tell me the story of two brothers he knew that did a lot of fencing: one of them got his hand mangled by the other. As one brother pulled the hammer high above his head, the other was testing the fence post for firmness by putting his hand on the top and giving it a shake.

Granddad would then roll out the fencing wire ('pig netting' he called it) and then fold it in half, creating a kind of crude wire envelope. Then I would hold it to the posts and he would tack on the wire so that it was nailed on at chest height. When we had finished, it looked like some half-baked fishing net. But then we would cut open half a dozen hay bales. Good summer hay. The most beautiful smell on our farm, a smell that is sweet and good. It is a breath of sunshine in your face in the depths of winter. It breaks open in thick slices that reveal the pressed flowers, vetches, grasses and herbs that were folded into it by the baler in July. As we spread it in the winter for the ewes, the ground is scattered with countless seeds of meadow grasses – Timothy, Common Bent, Meadow Fescue, Yellow rattle. Some farmers feed more generously than others; others see feeding as a weakness in hardy mountain ewes.

The truth is, we feed our ewes well with hay, and try and hold their condition as long as we can to ensure strong healthy lambs. A week ago, the ewes would have turned away from this hay, because fresh grass was still available, but now they line up and start to tug handfuls from the hayrack we have built. We stuff the encyclopaedia-thick slices into the wire envelopes that thread away across the hillside.

For the last ten years, my father and I have put out by hand a ton or two of hay each morning in about a dozen different racks and other contraptions across the farm to keep it accessible and as dry as possible until it is eaten. Working on a frosty morning it can be idyllic, but there are not so many of those days. Usually it is wet or cold, and your eyes ache with hayseeds blown into them. We slather, slip and slide about in the mud that is created around the hayrack. The wind catches the lids of the hayracks and threatens to hurl them away, or rips a gate from your hands and clatters it back against the wall. There is not enough daylight to get all the outside work done. When urban friends come to visit, I get uncomfortable as they drink tea and chat at 3 p.m., because I know (and they don't) that that last hour of daylight is all I have to do three more jobs that are impossible in the dark. In some ways, this makes you an uptight pain in the ass. But it speaks of the fact that electricity freed most people from the cycle of the sun in northern climes. There is no light switch for our land so we live by the up and down of the sun.

These are the days when the wind blows right through you, filling you with a sense of hopelessness. Days when the sheep stand sourly behind the walls. Short, sullen, dark days in winter when you are just holding on; days when you can hardly stand up and you can't help but be aware that man is a feeble thing in a hostile universe that doesn't care.

We dwell, it seems, for weeks in midwinter inside a grey cloud. Everything is wet to the touch, and slowly rotting back to the earth. Deep-green moss half-hides the curve of stones in the walls like the quilt over my children's entangled legs each morning when I leave. Silver lichen clutches out to the air from gate stoops, branches and fence posts. They say we grow lichen here because the air is clean. Uncorrupted. And sometimes you can taste the sea salt in our winds, though we live an hour's drive from the Irish Sea. The land becomes sodden. Fields running with water, bubbling from drains and springs untamed. The hillsides seem more water than land sometimes. Men and sheep wear out fast here. We beat the winter by still being here when it blows out, and by recovering quickly in the summer. Sometimes I think our sense of belonging relates to how much weather we have endured – we belong here because the wind, rain, hail, snow, mud and storms couldn't shift us.

Our resistance to change is the key to us. Through my life my father has resisted each new technology that professes to change everyone's lives for the better. Quad bikes, mobile phones, credit cards, computers ... each in turn was met with complete scepticism and years of resistance.

I won't lie and say I love each day of winter, because I don't. But there is the dream of summer to carry me through, and moments of beauty that transcend the mire and the slog. Snipe bursting from the sieves as we approach, and hares watching, then bursting from their well-worn forms at the last moment. The daylight wanes halfheartedly. Flocks of fieldfares fold backwards under wings flashing silver, tumbling over the wind and away down the thorn dykes.

The little becks that fall off the fells through our land can mostly be covered by a man's stride. They fall, wriggling through the rocks, down the rocky hillsides, little more than a trickle at first, but soon becoming frothing white ribbons in a few hundred metres, connecting us to the Irish Sea and the Atlantic.

My grandfather always waited for the spates of late November or December because he knew what it would bring. A harvest. Salmon. Sea trout. He used to walk the becks each day whilst shepherding and would come home in the early winter excited that he'd seen a flash of silver, a muscled bow-wave riving upstream. The fish were back.

They say that local lads used to poach these fish, working in gangs at night with torches to illuminate the riverbed, metal gaffs to snag the fish on, up to their waists in cold beck water. They say that it was exciting, that the shiver of a strong salmon on your fork prongs was a thing to make you feel alive, that the fun was in being abroad in the small hours, fearing having to fight, or outrun, the river bailiffs sometimes, all under the noses of the incomers, who at any moment might glance out of their windows and see strange torches flashing in the valley bottom, and call the police. They say it was fun shouting out to mates that you'd gaffed one and flung it into the rushes. They say all this, but I know nothing. Poaching is illegal. The seas have been emptied of fish by industrial-scale fishing. But we still find gaffed fish lying forgotten, lost, on the gravel sometimes, and see a flash of silver in the shallows.

One brilliant blue frosty morning we heard hounds to the west, singing of a fox. I was standing with my grandfather as he fed the sheep. The hunt was hidden from our sight by the brackened contours of the fell, a couple of miles away.

We often watch the hounds working around us, or passing over our land. The fell packs aren't the red coat affairs of the Home Counties. This is simply working men on foot following a pack of wily fell hounds after foxes that often get the better of them in the rugged terrain. Some keen-eyed old men with binoculars follow from their cars on the nearest roads. Whatever the rights and wrongs of it, it was a spectacle in winter. My granddad was no animal rights activist, but like many of the local men he had a begrudging respect for foxes, or 'Reynard' as they called him rather grandly. To us, foxes are not things to be pitied, but tough and wily creatures quite capable of looking after themselves.

We often used to watch the foxes giving

the hunt the slip. Granddad would have a wicked smile on his face as he watched. He seemed to support the underdog, the fox, except at lambing time when he'd lost lambs and then he was fine with the fox getting his come-uppance.

A fox catches our eye back away to the east at the far end of our land, a little, bright red dot in the sunshine, where the noise of the hounds had led our eyes. He is sloping across the ground with that effortless gait that carries foxes across miles of land seemingly without much effort. The sun catches his coat so that he seems to glow a fiery orange. He purposefully cuts through every hole in the hedges, under every gate bottom, making his way across the fellside he knew so well towards us. A way back, a mile or so behind, the lead hounds were on his scent. They, too, catch the sunshine and glow white like pieces of fine china tumbling down a hill.

The fox crosses the road and enters our field. He lopes directly towards us. I step closer to my granddad, and his fingers clutch my shoulder excitedly. His grip says 'Hold tight, watch this'. The fox had seen the sheep we were feeding and wants to lose his scent in their smell. He darts through them, less than fifteen feet from where we stand. The sheep don't seem bothered, but part to let him through. He pauses for a step and casts us a look. Then he circles the ewes and glances back across the fields to the hounds, three fields away. Just as if he is judging time. Then he shoots down the bank behind us and across into the rushes of the boggy valley bottom. We see him making his way through the sieves.

Now the hounds are just a field away from us, their blood up, hot on the scent. But they don't know the land like the fox. They miss the holes in the fences and the gaps under gates and are bunching up, or jumping. They lose ground all the time to find the scent again. My heart is pumping out of my chest. The lead hound gallops up the field towards us and the ewes scatter away to the quiet end of the field. Other hounds jump the wall and follow. The rest of the pack can be seen making their way to us from fields back. The lead hound casts us a pleading look, and we sort of shrug, amused. He raises his nose to the air, trying to separate sheep and fox smells. The following hounds join in to circle us, confused. Then one of them catches the smell of the fox at the dyke where he'd exited the field. The hound song starts again, and they tumble through and over the fence, heading down into the bottoms.

They never did catch the fox. We stood and watched them try to navigate the confusing scents in the bog. We saw five foxes that morning, heading in different directions from the valley bottom. The hounds seemed baffled. My granddad smiled, and said, 'Them clever bloody foxes are running circles round them hounds.'

Two ewes have gone off their feet. Slavering at their mouths, shaking, unable to stand up, they are hunched pathetically in a heap. I saw they were poorly when they were unable to run up to the feed when I fed their flock. One has its head on its side and looks to me like a case of listeria, a disease that affects the brain, strikes suddenly, and usually ends in death, despite our treating them with anti-biotics. I lost a good sheep a week ago to listeria and this looks like more of the same. I catch them and take them into the barn. I fill them with drugs and go for a coffee in a sulk. I leave my father standing silently, watching them. I am in a foul mood, because these are three of my most well-bred females. Something doesn't add up, but I can't work out what. Listeria doesn't usually drop sheep off their feet in quite the same way as this. Something about these cases seems related to the weather suddenly going colder. But I am too fed up to think straight. Half an hour later, my father passes judgement: 'That's not listeria. Them sheep has staggers. I've given them a shot of calcium and they're a bit bet-ter already.'

He was right. There is always someone who knows more than you about sheep, usually someone older. Staggers is a condition caused by calcium deficiency. Sudden changes in the weather or growth of the grass can bring it on. It is more common in older sheep when the first flush of grass comes. Yet these young ewes have it. The cure is simple. You inject a large amount of diluted calcium fluid into them, under their skin, and then stand back. The prognosis is much better than for listeria. Sometimes they get up and go straight away. An hour later, these two are still in a bad way, but you can tell that whatever was troubling them has eased. Good stockmen spend a lot of time looking, watching and thinking. That's what they are doing when they seem to be standing doing nothing, looking over a gate as you pass them on the road.

My dad has been very cool about my double life. In fact he encourages it. We will be working in the sheep pens as two equals, busily grafting away, and then he will suddenly stop, look at me, and say, 'Isn't there something you should be doing on your computer? I can do this on my own.'

Don't get me wrong, my dad and I are still all about the sheep. He knows sheep are my obsession, and that, given half a chance, I would do nothing else, but he also knows that you need to do other things here to make a living, and that this has always been true.

Though we have the chance to keep our farm and way of life going only because of the other things we do, it takes my father, my mother, my wife, my kids and the wider family to make this farm what it is. The sensible thing is to use each of us to best effect. So if there is the faintest hint that I should be doing something else that would bring more money in, then I am banished as quickly as possible to do it.

I used to hate these tensions, this being pulled in two ways at once. It went against the feeling I was brought up with that the farm should always come first. But I've grown used to it. Part of that is that I see many families like ours all finding ways to have one foot in the modern world and one in their living past. Many of my farming friends run campsites or B&Bs, their wives working off the farm to support it, or working off the farm themselves seasonally. This is how crofting families work in Scotland, or how families survive on their farms in places like Norway.

I have been to many places where the past traditions have disappeared and the people regret it. In the valleys of Norway, they are trying to encourage people to farm again in some places because the character of those places alters without it. Farming is more than the effect on the landscape: it sustains the local food industry, supports tourism, and gives people an income in places that might otherwise be abandoned. In some remote areas of Norway, it is difficult to manage forest fires without the remote farmsteads being there to watch and raise the alarm. But above all, when local traditional farming systems disappear, communities become more and more reliant upon industrial commodity food products being transported long distances to them, with all the environmental cost (and cultural disconnection from the land) that entails. They begin to lose the traditional skills that made those places habitable in the first place, making them vulnerable in a future that may not be the same as the present. No one who works in this landscape romanticizes wilderness.

Like much of the rest of my life, my marriage to Helen has followed a curiously old-fashioned pattern. She is from exactly the same background as me, a family farm in the Eden Valley. Her father kept a herd of dairy cattle and a flock of sheep. Long before I knew Helen, I sold sheep to her father and was on first-name terms with him. When I first appeared at his house in my best clothes to collect Helen for a date, we spent ten minutes discussing the price of sheep (much to Helen's acute embarrassment and annoyance).

Her father is a friend of my father. My dad had been round there and got drunk after a sheep sale some years earlier, been sick in the bathroom when he was worse for wear, blotting our copybook with my future mother-in-law. She took a little convincing, apparently, that I was a suitable catch. Helen's grandfather had bred some of the best Clydesdale horses in the country. He had been a friend of my grandfather. For generations back it goes on like that. The same characters appear in their family stories as appear in mine. Our grandmothers were lifelong friends, so much so we speculated whether we had been set up. When I was first introduced to Annie, her grandmother, as Helen's 'boyfriend', she told me she had ridden to a dance on the back of my (great-) uncle Jack's motorbike. She smiled at the memory, and I couldn't resist asking if he was 'fast'. She giggled, recognizing my innuendo, and said, 'Aye, in more ways than one.'

Uncle Jack, or 'Peo' as he was universally known, was a famous character in our area. He'd been a farmer, racehorse trainer, egg dealer, and God only knows what else in his lifetime. When my father was young and fresh from passing his driving test, he would be sent to drive Jack somewhere. It would inevitably turn into a boozy session in some pub or farmhouse miles from home, with my dad delivering everyone home to farm-houses around the county in the early hours. Jack always had rolls of 'egg money' in his pockets, dealing in cash with local hotels (so the taxman never caught up with him). The thick wads of notes would flop out of his pockets like he was some Sicilian mobster, with him seeming to think it was the most natural thing in the world.

Once he was walking cattle to the auction with my father. An impatient young man, probably heading to work in the local town, was behind the bullocks on the road in his new Mini. The man drove right behind the bullocks, stressing them, revving the engine and providing a running commentary about how late he was being made. Jack told him to settle down, but he kept fretting and com-menting on how slow things were going while pushing too close to the cattle. Then one of the bullocks turned suddenly and belly-flopped over his bonnet, leaving a cow-sized dent in the bodywork. The young man jumped out, whining and complaining and throwing his hands up in horror at what had happened to his beloved car. The men driving the cattle thought he had got what his pushing deserved, and kept walking. But Jack turned back, prodded the man to stop him talking, and asked him how much his car was worth to buy. The young man said the price. Then, without so much as a haggle, Jack peeled off the price in fresh £50 notes from his wad of egg money, stuffed it into the young man's top pocket, told him to park what was now 'his car' in the lay-by, and to 'fuck off and stop pestering folk'.

When I knew him, he was an old man and would sit at my grandmother's table and suck on boiled sweets that she placed there for his weekly visits. He was famous, amongst other things, for arranging and being guest of honour at his own 'wake'. They say he had invited hundreds of friends to the mother-of-all-parties at a local hotel, long before he had any form of illness, let alone a sense of impending death, just because he reckoned his wake would be a lot of fun, and he didn't want to miss it. A few years later, and still in rude health, he organized another and invited everyone again. You can say his name to just about anyone over fifty in Cumbria and they have their own Jack Pearson story.

It is the week before Christmas and my elder daughter is holding a sheepdog puppy. These two things should not be connected, but I fear they are about to be. If there were a prize for the world's cutest puppy, this one would walk it. It is a black and white bitch. We are standing in an old barn belonging to a good family friend called Paul. He breeds fine working sheepdogs and from time to time sells one or two that he can spare. A good sheepdog that has been trained is worth thousands of pounds, so the good families are held on to and getting a pup from one is difficult. It has taken us a few years to get the chance of a puppy from him. He loves his dogs and clearly hates seeing them go to an owner that might 'waste them'. That he's letting us choose one is a bit of a privilege.

I know we will only get one chance. Make a mess of training this pup and we won't be offered another one in the future. My daughter is inspecting my face for signs of weakness. Her look tells me she is searching for a 'Yes'. We haven't been told which one we are allowed to take yet, so she may be disappointed.

But Paul knew what he was doing when he handed the puppy to her. He smiles and says that she is the last bitch that hasn't been claimed. I am pleased, but my daughter looks as if she wants to get in the car quickly before he changes his mind and withdraws the offer. I almost have to prise the pup off her when we get home. She wants to take it to bed, but I have to tell her that a sheepdog is not a pet.

It is easy to waste a sheepdog. I know because I did it when I was twelve years old. Dad allowed me to keep a fine-looking puppy called Laddie. I didn't know how to train him properly and got frustrated when he couldn't do what I wanted him to do. I'd raise my voice and he would become confused or scared. It was a bad combination of a young dog that needed instruction and me not knowing enough to guide him. Fewer farmers than you might imagine know how to train a dog, or have the time to commit to doing it well, so many dogs can do the basics but little more. It is difficult to get a sheepdog to work well and understand you. It required more wisdom, patience and kindness than I had. It still tests me.

Laddie was a useful dog on the farm for the next few years, and there were moments when he did good things and we understood each other. Once, we sorted two ewes that we needed for a show off a hundred others we didn't need in a field and walked them home. But it was a rare moment, and I always knew he wasn't as good as he should have been. Sometimes he'd run home when I'd lost my temper and shouted at him. He lost trust in me. I knew whose fault it was. I knew that I'd let him down. I look back and think he would have made a good dog if I had known a bit more. But a man's life comes full circle; you can learn and do better than in your past. I am determined not to make the same mistakes again. We called the puppy Floss.

Floss learned fast. I tried to have two short training sessions a day with her, starting by teaching her to lie down, walk to heel, and come back to me when loose. Then I introduced her to sheep. She was unsure of them to start with, but when they raced away from her, she couldn't help herself: something in her body took her whizzing past them and holding them back to me. We kept at it in little bursts, building up her confidence, until she could hold these half dozen sheep whichever way I let them run past me. After little more than ten days, she was working like a sheepdog. I built a round pen for the sheep, and let her run around the outside of them. I encouraged her with commands so that when she went clockwise, I gave the command 'Come bye', and when she ran around anti-clockwise, I said 'Away'. Then we took this to the field and she got it immediately. There was a thread of understanding between us, but it could break at any moment. When training a young dog, it breaks all the time – snap – then she is confused, frustrated, lost. The training is all about finding that thread of connection, finding that understanding, trust and belief in each other.

Some shepherds are wizards at training dogs. I am an amateur, so I rang Paul and asked him questions and he patiently shared his knowledge. I started to think he had sold me a great dog. She was timid when not working; like many sheepdogs, she didn't want to be a pet, she was all about the work. Floss made me look good – you could show her once how to do a thing and then she knew. She got faster, stronger, fitter. She listened intently. She wanted to know what I wanted her to do before I needed her to do it. She turned almost before the command syllables were uttered. This is about more than command and response. It is more like a shared understanding, a shared thought. She is an extension of my brain and my arms.

But she was still green then, and would do a thing that she thought needed doing whether I wanted it done or not, like holding sheep from going through a gate I wanted them through. I just about stopped myself raising my voice and sounding mad. I called her back and showed her what I meant. She almost smiled as she came back to my feet. I felt blessed to have a dog that can work like this.

One day, my sister and her husband came to help my father on the farm. They made a schoolboy error and drove his quad bike up a field without realizing that the bags of sheep cake (a grain mix concentrate) in the rear had toppled over and were being spread over the fields, wasted. When they got back to him, oblivious to their mistake, he exploded (insert your own swearwords). My brother-in-law, who is a mild-mannered and kind man, and slow to anger, was furious at being spoken to like that, and stormed off with my sister in tow. When they passed me down back at the house, as they got into their car, he turned to me and said, 'Your dad is a fucking loose cannon.'

A day or two later, this had, like most family rows, blown over, but the name 'Loose Cannon' has stuck. It has become Dad's nickname in the family. Even he smiles. Helping my dad has always been a risky business. You can quite easily end up falling out with him. Once, when I was back for a weekend from university, I got out of the car and he stormed past me, cursing and swearing, clearly not pleased to see me, and clearly fighting some other losing battle. I wasn't in the mood after hours on the road. I just shouted after him, 'Should I fuck off and leave you to it?' And he replied, 'Yes, fuck off.'

I got straight back in the car and went elsewhere. Some battles are best avoided.

Dad would disappear off to the local auction to buy us a turkey. The little rural auction markets are the clearing-houses in the days before Christmas for any table-ready birds that are not sold direct to people on the farms. Often there is a last-minute glut and bargains can be had. As he drove off, we would turn to each other and smirk, because Dad never just buys a turkey. The bargains on offer are sometimes too much for him to refuse. He loves bidding for things, seeing that they make the 'right price'. Sometimes he gets the turkey he was sent for, but usually he comes back with enough assorted poultry and fowl to put on a medieval banquet. It all depends on the 'trade' (prices). If the trade is bad, he will not be able to help himself and he will fill the car up.

He arrives back later that night, beaming at what he has done. My mother goes out to the car and comes back shaking her head, asking what the hell is she meant to do with all those birds. She asks him what anyone is actually meant to do with six turkeys, three geese and a partridge, minus a pear tree. Dad shrugs like it isn't his problem (he is confused why women are always so negative). It is good cheap meat, he says, half the price most folk had paid for their Christmas dinner. We can freeze it and have it in January. My mother groans, and reminds him that the freezer is still full from his 'bargains' of last year. We all laugh, and everyone agrees that letting Dad go to a turkey sale is a bad idea. We tuck into cold turkey in July with some chips and tease him when he says it is 'a bit dry'. We laugh and tell him next year we'll stop him going to the turkey sales. But, of course, we never do.

My younger daughter's eyes are so wide with excitement she looks like she might explode.

'Dad, wake up ... he's been.'

'Eh? Who?'

'Father Christmas!'

'No way!'

'Yes way... Dad, I've got a stockingful of presents.'

The pattern of our Christmas is the same as it was when I was a child. It has always been the tradition that the children can open their 'stocking presents' from Father Christmas when they wake up (as long as it isn't too early). So they pile into our bed. Then, in a frenetic blur of ripping paper, Christmas begins. Soon the bed is littered with crumpled wrapping paper, sticky tape and kids stuffing sugary sweets in their mouths. Once the stocking presents have been opened, I go outside to feed the sheep, Helen puts the turkey in the oven, and the children have to sit on their hands, enjoy what was in their stockings, and bide themselves until I get back. They are not allowed to touch their main presents under the Christmas tree until the sheep are fed and I come back in and have had my breakfast. I'm not sure how long it has been like that in our family, but the lesson is a simple one. The farm and the live-stock, and the men and women that work, come first.

We work through Christmas. The sheep need feeding and looking after as if it is any other day. It sounds like a pain, but it isn't. Tending to a flock of sheep or feeding cattle feels like the most natural thing to do on the birthday of someone born in a manger in a faraway land of shepherds. We go to church the night before and see our friends and neighbours, and I enjoy singing those carols that are about shepherds, and eating mince pies. We can make things easier by readying things on Christmas Eve, so the day before is fairly hectic. Hayracks are filled, feed bagged ready for the morning, pens mucked out, and a host of little routine jobs done so they don't need doing on Christmas Day. Then it is just the core shepherding work that is needed, each batch of ewes to feed and check they are OK. It is kind of nice to be outside doing something decent on Christmas morning. I see cars flowing down the wet grey roads. People heading to their relatives to see presents unwrapped. My neighbours pass by and wave on their way to their sheep in the fields around the valley and beyond, bales of hay and bags of cake stuffed in the back of their vehicles. When the sheep are fed, we can sit down and enjoy a day of presents and gluttony.

Like my father before me, I make the kids endure a long wait as I shepherd the ewes. When I get in, the kids quickly serve me a boiled egg and some toast and plead that I hurry up and eat it. On no other day of the year do they pay any attention to my breakfast. Our youngest, Isaac, comes to tell me that the living-room sofa has presents on it for him. He is desperate that I come so they can open them. I give in, and we enjoy our Christmas as a family. Isaac gets a lot of books about farm animals, some toy sheep for his farm, and some games. The toy sheep are his favourite thing. He 'shows them' like he is trying to copy someone else he has seen doing it for real. It seems a serious business. He tells me he needs a sheepdog now, like Floss, and then he can come and help me and his granddad. I ruffle his hair and tell him he can borrow my sheepdog for a bit longer, and that maybe Floss will have pups one day. Then I tell him there is more to life than sheep and sheepdogs, but he looks back as if I've just said something idiotic.

Presents are unwrapped, chocolates munched, a giant turkey dinner with all the trimmings devoured. The Queen's Christmas address is watched and the National Anthem emotionally hummed to. Then I usher the kids out to get some fresh air, which none of them want, but they are all nicer for it. They all have to do some jobs every day and Christmas is no different. This way they learn about duties and responsibilities. Working makes the food and family times later in the day more meaningful: we have earned the rest through work. I'd hate not to work at Christmas.

My children have long figured out what makes me tick. When my elder daughter was four years old, she looked at me sternly across the kitchen table and said, with a wisdom beyond her years, 'The trouble with you, Dad, is that it is all about the sheep.'

My father has cancer, but somehow he has managed to come home from hospital in Newcastle for Christmas. Though very ill, he insists on coming to the farm for his Christmas dinner. He is an ashen-green colour and has to disappear to the toilet every few minutes. But he eats some dinner. The women of the family bustle about, making Christmas happen. The kids play on the floor.

Dad is happy and relieved to be home in his place in the world. From our window, he can see his tough, wind-battered, rain-sodden farm stretched beneath us in a rare moment of winter sunshine. His tear-reddened eyes soak it in, as if it might be the last time he sees it. 'Look how that new tup stands.'

The stock tups are grazing away across the hillside and have caught his eye. One swaggers towards some ewes penned the other side of a wire fence. We don't need many words. The tup was part of the story of the autumn past, when we had fallen for him and taken a chance, buying him for a high price. He is the future of our flock. Our choice. I'd nudged Dad's elbow and egged him on to buy it when he had wavered at the auction. He'd loved that. The sons and daughters of this tup will be born in April and sold the next autumn. We buy dreams of the future when we buy a tup like that.

The next two-year cycle in the life of the farm plays out in our heads. I catch my father's eye and his look tells me he loves these things, but that he might never see them happen. It also says, he knows this will all go on. I turn away to cry so he can't see.

It is January, and I am wading through a grey swell of sheep. They are in their midwinter coats and the wind buffets their backs so the wool ripples in little waves. The oldest ones thrust their heads against my legs, or towards the hessian feedbag, eager to get at the sheep cake. I stumble on through them, looking for a clean bit of field to feed them on. For a few short weeks until lambing in April, those ewes scanned with more than one lamb in them will be fed some sheep cake. I carry this out in sacks, as cobs, little cylindrical nuggets, and pour them in a line on the ground. Momentum is everything, or I will find myself wrestled to the ground, the feed spilled, and unceremoniously mobbed. As their pregnancies progress, we have to be vigilant for any issues.

275

The weeks from Christmas until March are the longest and most testing of the year. I finish work each day in the gloom, orange specks of light across the valley telling me my neighbours are still working as well, although on bad days you can't see the other side of the valley and I'm greeted by waves of rain or snow that pass by as if in slow motion. These hard, cold, wet weeks are when Herdwicks come into their own. Few other breeds would survive the winter here, carrying lambs in their bellies. The ewes are becoming heavier in lamb with each passing week. They need our help. The bond between shepherd and flock is formed in these cruel months.

I missed large chunks of a couple of winters while I was at university. I missed the feeling that would come in the spring, that elated feeling, because I hadn't earned it through the hard months of winter, and the sun didn't shine quite as bright, or the grass look so green.

I understand why people once worshipped the sun and had countless festivals to celebrate spring and the end of winter. It is this endurance of everything that nature throws at us, year in year out, that shapes our relationship with this place. We are weathered like the mountain ash trees that grow here. They bend away from the wind, and are battered, torn and twisted. But they survive here, through it all, and they belong here because of it. That weathering makes us what we are.

So you live for those little signs that you've outlasted it, the point when the days lengthen in March or April and eventually warm up, the fields turn marginally greener, and the sheep suddenly lose interest in hay as the grass begins to grow. Grass is everything. We see a thousand shades of green, like the Inuit see different kinds of snow.

These are the ways that winter shows it is passing: the creeping out of the daylight each day, the warmth of the sun increasing, the bite of the wind easing, the grass greening. But the ravens honking above the fells speak of carrion from worn-out ewes, and the field-fares flashing out of the hedges are reminders that winter still holds the far North. Foxes steal withered-up moles from the barbed wire where the mole-catcher has left them, telling of the hunger that once would have tested men here as well as animals. The carrion crows still lord it over the valley, cawing from the tops of thorn bushes or trees. We know that without warning winter can grab hold of the land again.

My father says you cannot rely on winter being finished here until May. Sometimes I think he is too pessimistic; sometimes it is done halfway through April. Late in the winter, things start slowly to change. Skeins of geese pass over, sometimes low and loud, other times high up near the clouds, heard as a quiet child-like chatter. We clear sheep from some fields in the first weeks of the New Year, so they can freshen up with spring grass for the ewes and lambs coming back to them in early April, and we look nervously at the pile of 'big bales' of hay, because the heap shrinks every day through the cold weeks and starts to look as if it will run out. My grandfather used to pick up little wisps of hay in the summer as if they were worth a fortune – 'that'll fill an old ewe in the depth of winter,' he'd say. Most years, it lasts into April with a little bit left, then the grass grows and you curse having any left at all.

By March we have to be vigilant for any issues with the pregnant ewes. We scan them in January so we know which ewes have single lambs in them (these ewes can, by and large, look after themselves with some hay) and which have twins or triplets (these ones need more help and are at risk of twin lamb disease or just getting worn down, so need feed and hay). The scanning is a blur of activity as our friend who does it gets paid by the sheep, so understandably wants the process to be efficient. We have to be organized.

He glares at the fuzzy grey screen, with one hand under the ewe's belly, shouts out 'Single', 'Twin', 'Triplet' or 'Geld' (barren) and a mark is quickly sprayed on to the sheep's back. She is then released back to the flock standing in the yard. Our mountain ewes average about a lamb and a quarter to a lamb and a half each. Any more and we get too many problems, as a young hill ewe can manage one lamb but two can be too much on marginal land. We don't want triplets as it inevitably means little lambs that are at high risk in bad weather or that need to be taken off the mother at birth so that she has enough milk to rear the remaining ones. Our farming system is not about maximizing productivity, but producing what we can sustainably from the landscape.

It took traditional communities often thousands of years to learn by trial and error how to live and farm within the constraints of tough environments like ours. It would be foolish to forget these lessons or allow the knowledge to fall out of use. In a future without fossil fuels, and with a changing climate, we may need these things again.

My other work has taken me to historic landscapes around the world, including those that face similar challenges to our own. I have met and talked with hundreds of farmers, stood in their fields and their homes, talked to them about how they see the world and why they do what they do. I have seen the tourism market shift over the last ten years with greater value attached to the culture of places, seen people growing sick of plastic phoniness and genuinely wanting to experience places and people that do different things, believe different things and eat different things. I see how bored we have grown of ourselves in the modern Western world and how people can fight back and shape their futures using their history as an advantage, not an obligation. All of this has made me believe more strongly, not less, in our farming way of life and why it matters in the Lake District.

Now, when I look at the world, and wonder if we will survive in it, I am full of hope for the future. There are young people coming into the farming way of life here. I see the pride in their eyes and their tough northern love of this place and our culture. This way of life continues because people want it to. If they didn't, it would already have died. It will change and adapt, as I and others have, juggling it with more modern lives, but the heart of it will remain. I now believe we will survive doing what we do. And, like Wordsworth, I believe our way of life represents something of wider benefit that others can enjoy, experience and learn about.

The choice for our wider society is not whether we farm, but how we farm. Do we want a countryside that is entirely shaped by industrial-scale, cheap food production with some little islands of wilderness dotted in amongst it, or do we, at least in some places, also value the traditional landscape as shaped by traditional family farms?

Recently I was in the south of China, following a winding path down the steep sides of a valley. As we got halfway down the hillside, we came across a lady under a tarpaulin tent, selling souvenirs. The things for sale were nice (though not made locally): little ornaments showing the villages I had come to visit. She flashed me a smile, but there was something in that smile I didn't quite trust. It was a plastic, have-a-nice-day kind of smile. I asked my interpreter to ask the lady if she liked selling these things, and she said she did, that she is doing very well financially. Then I asked what her family did here before tourism, and she said they were duck and pig people. They had farmed ducks and sold meat and eggs, and fattened pigs for centuries.

I told her that when I am at home I am a farmer, and she smiled, but this time a real smile, open and friendly. It vanished when I asked if they still farmed ducks and pigs. No, those days were past. Someone had decided that ducks and pigs were too messy, they make too much shit, and tourists don't want shit on their shoes.

Like so many of the loved places in the world, this one is struggling to cope with the tension between wanting to earn money from tourism and its potential to sweep away that which is special in the first place. Walk enough people over a stone step and you will eventually wear it away to nothing. So, someone had decided that keeping ducks and pigs is yesterday's work, and that selling souvenirs is today's. When I asked her which was better, farming or selling souvenirs, she told me there was more money in souvenirs, but she'd rather keep ducks and pigs, because that is what made her family and these villages what they are. Later, as I walked through those villages, I had to admire how clean and well preserved they were. But I looked at my clean shoes and I felt this was a bit of a sham.

My shoes should be mucky.

Spring

Let no one say the past is dead.
The past is all about us and within.
Haunted by tribal memories, I know
This little now, this accidental present

Is not the all of me, whose long making
Is so much of the past.
...
Let none tell me the past is wholly gone.
Now is so small a part of time, so small a part
Of all the race years that have moulded me.

Oodgeroo Noonuccal, 'The Past', from
The Dawn is at Hand (1992)

And yet all these impressions, and a thousand
more, add up only to a one-sided, personal and
entirely superficial memory, which ignores what
the mountain may mean to those who have lived
for years beside it.

Norman Nicholson, *The Lakers* (1955)

I respectfully maintain that work, business and
the undisturbed customary use of centuries
should be set before idle amusement.

Mrs Heelis (Beatrix Potter)
in a letter to *The Times*,
January 1912, objecting to an aeroplane
factory on the shores of Windermere

My second daughter, Bea, has come to work
with me in the lambing fields and has seen a
ewe lying against a wall, down the field, paining on its side. It is lambing. She has come to
lamb a sheep, determined to keep up with

283

her elder sister, who lambed one two days ago and is crowing about it. She appeared in her work clothes as I was leaving the house at first light, and got on the quad bike with me. I tell her that it is cold, and that she may not get a chance, but she seemed to know that she would and came anyway.

As I shout the ewes up to their feed, I see three pairs of large round ears appear out of a sea of sun-bleached sieves. The roe deer know my voice, and that I offer no threat to them, so they hold tight each morning and watch, unless I approach them. Bea and I drive down the field, the roe deer bounding off in half-hearted leaps. Floss pushes behind us in excitement because she knows I may need her to hold this ewe until I have caught it. I tell her 'Steady', because I do not want to jump off the quad bike and stress the ewe if I can help it. The ewe is fairly oblivious to us arriving and in three steps I catch her and hold her down before she can struggle. She lies on her side. With each contraction, she raises her head and throws it back against herself. I can see two legs and a nose breaching – as it should be – so my daughter can do this.

She looks nervous, as if she is trying to be brave. She hates her elder sister doing any-thing that she hasn't, so there is a grit behind that nervous smile that says she will do this thing whether it is fun or not. She is small, just six years old, and the lamb coming (judging by its feet) is on the large side. But she grabs a lamb toe in each fist and pulls. I talk her through it. I tell her to give the ewe a moment between pulls. I can tell she is a little unsure whether it feels right, but with each contraction the legs slip out a little further. She gets a hand around each leg at the first joint of the foot. Then the nose is breaching. She wants to stop and let me do this, but I reassure her, tell her she can do it, and if she does, she will be able to go back and tell her mum and her sister she did it.

Then the smile toughens and she starts to pull. My daughter is tired now. She nearly stops when it resists her pressure at its hips, but she knows enough to pull it further and get it out now so it can breathe quickly. She slops it down in front of the mother, whose tongue is already manic in its determination to lick it dry. My daughter laughs because the ewe licks her bloody hands as she sets the lamb down. She stands over the lamb, which wriggles and shakes itself free of the afterbirth slime, her face a mixture of pride and awe. The sun is shining, so we will leave this lamb and its mother.

Then my daughter remembers something. 'We have to go for breakfast, Dad, and tell Molly I lambed one. And it's bigger than the one she lambed.'

I used to follow my father (and grandfather

before that) around the lambing fields. Now my children follow my father around the farm in all seasons and learn from him. He comes to the farm each day, he teaches my children his values and his knowledge just like my grandfather taught me. My son worships him. The thing comes full circle.

But this lambing time he is not here.

He is in Newcastle hospital where they are trying to kill the cancer with chemotherapy. I don't know whether they will beat it or not, but when you love someone, you have to believe they might.

Once, I wanted to kill him, and now I want him to live more than anything. He has lived his life by his values, a modest, hard-working life that I admire. But now he is in hospital and I can do nothing to help him. So I send him commentary and pictures of his best ewes and their new lambs on my phone so he can live this time vicariously through me. I am forbidden to go and visit, because he wants the farm looked after right more than he wants visitors.

I lamb the sheep alone this spring, but it is not the same. Dad couldn't care less about money and would drive twenty miles without a thought to help a friend with work if he thought he was needed. My grandfather might have bought this farm, but it is my father who has tended it. He is stoic, but we know it is killing him to be away from his sheep and from the work. Recently he became a pensioner and someone rang to ask if he would stop working. He just laughed at the idea, as if it wasn't even a serious question.

In the weeks that follow, my father recovers. He is in remission and almost back to full strength. We dare to hope it is gone.

I am on edge. I can't settle. It is a week before the lambs are due to arrive, but I am ready. Fretting. I go round the ewes more often than necessary because I'm nervous. Helen tells me I am a fool; that in a month's time I will be worn out and the work will start soon enough, that I should save my energy for when the lambs actually arrive. I know she is right, but I still go anyway. So much can go wrong in the week or two before lambing. Ewes can suffer from conditions brought on by the stress of carrying lambs, like twin lamb disease. They have come through winter and now are tired but heavily pregnant. There are countless things to worry about, and I fret about all of them.

We start lambing at the beginning of April. In theory, this is the point at which winter becomes spring here, but sometimes winter isn't aware of our plans and the weather is still gruesome until well into lambing. Snow. Rain. Hail. Wind. Mud. We brought the ewes quietly into the lambing fields in the valley bottom a few days ago. These fields should now have some grass, after being cleared of last year's lambs, which were fattened and sold to butchers in February and March. But there isn't much grass, so we are praying for some warmth, begging for spring.

I feed the ewes in long lines on the hillsides. They line up behind me, with their heads down, like a massive scarf. I walk up either side of them and check their rear ends for signs. One ewe has blood on her woolly tail. But she isn't lambing. She has miscarried. My heart sinks. There is always a dread lurking behind our excitement each lambing time, a fear that something might go wrong. This is nature, not some cute movie. I call the vet. She tells me that there are a lot of failed pregnancies going around. It is probably a virus. Bad news: there isn't much I can do. The virus will have gone through them weeks ago, early in the pregnancy without any outward signs. They may all have it, or just a few. Those that have it will miscarry a week before the due date. I inject all the ewes with antibiotics to help them get over the disease. The vet isn't sure whether this will work or not, and doing anything to the heavily pregnant ewes can induce stress and create new problems, but it is the only weapon I have and we always try to do our best for the flock.

Half a dozen ewes abort in the next week. Some of the lambs breathe briefly, then die. I wake up feeling sick each morning.

It is not the loss of these half dozen lambs that upsets me, as sickening as that is, but the pressing fear that this could escalate. But thankfully it doesn't. With each passing day, the fear that the whole flock will have dead lambs passes, and it stops as soon as it started. We come through this thing. It starts to look like a minor setback. As the miscarriages are found, I skin the aborted lambs as people here have always done. I cut around the legs and the neck and then peel the skin off, leaving the head and legs black and the rest of the body naked. It isn't a very nice job, but it takes some skill to do it neatly. Ideally the dead lamb's skin forms a kind of jacket that can be put on to a live lamb without a mother. You soon learn to get over your squeamishness on a farm. My children watch all of this, and I let them, because it is real.

Blood was everywhere when I was a child. Sheep lambing, bloody hands, dehorned cattle, zombie-like, with spurting blood as they careered off across the yard before being let out in the fields for spring. Cow Caesareans, men with armfuls of guts and blood, then shoving it all back in and stitching it up. One night we calved a cow and the vet said, 'There is something wrong with this blood...

Christ, this cow is a haemophiliac.' It bled to death despite our best efforts to stop it, but we rescued the calf. My father's hands were always torn, scraped, grazed, kinned and cracked with sores or dried blood. He never bothered about it and would call skin 'bark'.

'You're cut, Dad.'
'Aw, it's nowt, just knocked a bit of bark off.'

We accept blood as normal as long as it stops flowing, and the skin scabs over and heals. In traditional communities, blood is part of day-to-day life, something even children are familiar with. Asian families still often slaughter a goat in full view of the whole family in the hallway. Middle-class English sensibilities suggest this is somehow in bad taste. But I grew up with blood. I like blood.

I would rather my children saw the blood and knew it was real than had a childish relationship with farming and food – everything in plastic packaging and everyone pretending it had never lived.

Everything and everyone is at times covered in shit or snot. It is just part of farm life. You learn to accept that you will get spattered in shit at times, or slaver, or afterbirth, or snot. That you will smell of your animals. You can always tell how alien someone is to our world by how terrified of the muck they look.

A friend over the hill from our farm lambs a
fortnight sooner than us and has some lambs
without mothers, so I go (six times) for a
lamb from her. Orphan lambs are shared like
this so that everyone makes best use of the
ewes with milk and the lambs. No one really
wants to rear a 'pet' lamb on a bottle. They
don't do as well in the end. The orphan
lambs are a different breed to the casualties,
but it doesn't matter. The mothers judge
them mostly on their smell at this age.

I take the lamb in my hands when I get it
home. The skin of the dead lamb forms a
tight-fitting waistcoat, with the orphan
lamb's legs poking through the leg holes and
its head through the neck. With legs and head
in, the skin cannot come off. I put the lamb
in a small pen with the grieving ewe and hold
my breath in the hope that she will mother it.
The ewe sees the lamb with the false skin on
and glares at it suspiciously. Then she sniffs at
it, and is confused. It smells just like the lamb
she had a short while ago. She turns around
in the pen a few times as it approaches her
hungrily. She is not convinced this is hers.
She gives it a head-butt or two to knock it
down, while she tries to reconcile the competing urge to mother something that smells
right with the suspicion that I have had
something to do with this situation. And

perhaps she knows her lamb is dead.

But these old shepherding tricks work, because all six of the ewes take to their orphan lambs. Some take them within five minutes, a couple take half a day, but all are back in the fields with a lamb after two days.

They have reared them well. I see them each morning when I shepherd the different flocks, and I smile. I still owe my neighbour for those orphan lambs months later, but we run a kind of unwritten ledger and we will return the favour another year. She passes my lambing fields several times a day, on the way to her own lambing ewes. If she sees anything that needs my attention, she will bawl down the fields to my house for me to go and sort it. And I do the same for her.

One of my best ewes is lambing. She has been lambing for nearly two hours, and I was unable to come back until now to help her. She is in a good place and the rain has stopped. I daren't leave her any longer, so with a flick of the hand Floss runs by her and I have her by the neck with my crook. I reach under her and pull her far leg so she gently rolls on to her side. Under her woolly tail I feel for lamb legs. A moment after catching the ewe, I have a bloody arm deep inside her. You never know what you will find. Sometimes it is a wriggling tangled mass of limbs inside. If there is just one leg, it means the other has been left behind and I will need to go in and fetch it forwards for a proper delivery. If there are no legs but a head, then the lamb can get stuck and die on exit, and I will need to push it back in and get the legs coming in front of the head, as they should be, like a diver. I have to guess whose limbs are whose by sticking my arm in and by touch alone. The ewe lies flat, lifting her head from time to time in contraction (my spare hand holding her down firmly and my leg holding hers from the ground like a wrestler). When I find the right combination of legs I get the first joint between my knuckles and start to pull. As my fist comes back into view I have two lamb legs between my fingers, followed, after some steady pressure, by the head. The nose, wrinkling with the pressure, shows – then breaches. Then the head peels out. I keep the pressure on as the body slithers out, then the pressure eases as the body follows in a gush of yellow afterbirth in a crumpled heap on the grass.

The lambs are born with a cough, splutter and a shake of the head. They don't breathe for a couple of seconds, so I tickle their nostrils with a piece of straw or grass. Then they shake and cough into life. The ewe's tongue instinctively licks the lambs dry. I turn them and can see that they are two 'gimmers' (ewe lambs). These are the future of the flock; they will live in the fells for most of their lives. I find myself talking to the ewe, telling her she has done well. She nuzzles them to their feet. Then, in minutes, on matchstick legs, they are stumbling towards the teats. Instinct tells them that they have one chance to live and it relies on getting hold and sucking.

The line between life and death is often paper-thin. It is essential that they get enough of the yellow creamy colostrum milk that carries the antibodies and nutrition they need in the first few hours. It is magic golden stuff, and half our work is ensuring that new lambs get to their feet and get their share. You rarely have to help a healthy Herdwick lamb on the second or third day if its mother is milking it properly, but a share of the Swaledales need a little assistance. When you skin a Herdwick lamb, you can see why they are so perfectly suited to this landscape and its tough weather. The fleece at birth is about half an inch thick, a leathery skin on the inside and a wiry weather-turning carpet on the outside. They are, literally, born dressed for a snowstorm or a rainy day.

A few years ago, we tried a tup from a modern lowland breed, a French type called a Charollais. When the ewes had lambed, it was snowing. The Herdwick lambs, at two days old, were racing against each other and skipping in the whiteout as if it was a sunny day. The French lambs of the same age were cowering and shivering behind the walls, and we had to lead them into the barn to keep them alive. I swore then that I was sticking with the proven native breeds on our farm.

I love lambing time. In the long, sodden and wind-lashed winter weeks, I sometimes daydream of escaping the muddy tedium, but I wouldn't want to miss it like my father is now. Every minute of your time counts. I've always loved it, ever since I used to follow my granddad around helping him feed the ewes in pens of little hay bales, and the pet lambs, sometimes being given one to lamb like my daughters do now. He would trail endlessly round from first light to darkness and afterwards, until I could not keep up and was sent to bed. He would check and double-check for emergencies, ewes or lambs in trouble.

I always marvel at how gentle some of the men were at this time of year, how you saw them kneeling in the mud or the straw of the pens, delicately threading a stomach tube down an ailing lamb's throat, over the little pink tongue. You could see how much they cared. My dad would be gutted if he lost a lamb; it would hang over him like a grey cloud, until he had put things right by saving others.

The arrival of lambs is on some kind of bell curve. It starts as a trickle of one or two each day, peaks a fortnight or three weeks later in a hectic blur of dozens, and eases off into a long tail of individuals over the following three weeks, until we eventually say enough is enough and leave the last few to lamb in the spring grass and sunshine without such regular supervision.

Lambing time has a kind of crazy daily rhythm. We are on a merry-go-round of responsibility, dictated by the need to get round the lambing sheep at regular intervals – every one to two hours. I know when I wake up that I will be at it for many hours. But I cannot predict what is going to happen on any given morning. I sometimes rush around the farm and do not see a single new lamb. Other times I find several new well-mothered and healthy lambs that don't need my help. The sun might be shining, and all can be well with the world. Or, frankly, it can be a complete disaster, with the shit right and properly hitting the fan.

I start work in the barn, a modern steel-frame building, before daylight, often with the previous day's ongoing troubles that still need help and the routine jobs like feeding the sheepdogs. The barn is like the maternity ward and A&E rolled into one. Because of the electric lights, I can work in there before dawn. Everything in the barn has 'issues' and needs the closer supervision or the shelter it offers.

As I step inside and turn on the light, the ewes in the pens call for their breakfasts. I rush around them with a hessian bag of feed to quiet them, and to prioritize problems. I soon see the most pressing cases. A young ewe has got herself bothered and turned on her own lamb that was born last night. She has lamed it, probably broken its leg. She is confused and may settle down later, or she may never mother it again, despite my efforts. I curse at her for being so mindlessly stupid and cruel. She tries to jump out of the hurdle pen and I wrestle her roughly back. The lamb needs a splint. I could jump in the farm truck and drive half an hour to the local vet, but that would mean I'd be away too long, and the lamb isn't worth enough. The vet will charge me several times the lamb's worth to mend it.

So I do what the vet would do: I've watched him plenty of times. I create a splint, padding on the inside and some strips of plastic to take the weight. It's a respectable job and has worked before. I capture the ewe's head into a head trap (like a medieval stocks), which gives the lamb a chance to steal milk for a few days, until she might start to mother the lamb again. She throws herself down in a sulk and nearly smothers the lamb that is displaying a complete lack of good sense by suckling at the wrong end, sucking a dirty piece of wool and not a teat. I curse at her again. Lambing time tests your patience and your goodness.

The length of time a newborn lamb can stay warm and alive varies a lot depending on the weather and how good a mother it has. A bad mother on a snowy or raining morning and it can be 'starved' (our word for frozen cold) in minutes, but with a good mother and on a sunny day it can be OK for two or three hours. My stress levels rise and fall in relation to that survival period.

Mountain sheep like ours are healthiest and most settled lambing outside, but that means I have a lot of ground to cover each day in the valley bottom fields. Many of our ewes lamb in the first two to three hours of daylight, so I need to get around every ewe on the farm as soon as possible after daybreak. I load the quad bike and trailer with feed each night, ready for the morning. Every minute's delay in getting around the pregnant ewes increases the chances of a disaster.

A lamb in a neighbouring pen, brought in last night because it had become separated from its mother in the field and got cold, has started with an ailment called 'watery mouth', and requires treatment. Lambs that get this start to slaver at the side of their mouth, and can be dead within an hour or two. I find the grey and red antibiotic pills that we treat them with and stick two over its little tongue with my first finger. The lamb gags and slavers them out again, and I curse and fumble about in the straw to find them. This time I push them over the top of the tongue and it swallows them down. Away at the end of the barn, an old ewe I had brought in days earlier, because she looked worn out, is lambing and doesn't have the strength to push it out. After a fight, I get two dead lambs out of her and she looks so sick she may die. I prick her with antibiotics, but I fear the worst. There is no romance in a morning like this. I haven't even got to the fields and most of the sheep.

The sun is only just rising over the edge of the fell.

By the time I get to the first field of lambing ewes, I am already wet. I look into the field and can tell immediately that all hell has let loose. The rain is biting cold and hillsides are just sheets of water. It is a disaster zone. A first-time ewe (a shearling) has dropped her lamb, when giving birth, into the beck, where it is stumbling and falling back into the shallow but deadly water. It is tough, but looks close to giving up, as it cannot climb up the bank. I lift it out and put it in the trailer. I send Floss to hold the ewe up, and after some slipping and sliding in the mud I have hold of her. I will take them home to shelter. The ewe looks uncertain of her lamb now, like the thread between them has broken. A hundred yards away, on either side of me, lie new lambs that look as if they are dead or dying. There is nowhere for even the experienced ewes to hide their newborn lambs from this downpour. Normally dry places behind walls are streams, sheltered spots have turned into ponds. The temperature is murderous. Cold. Wet. Windy. My neighbour says later this is the worst lambing weather she has ever experienced.

The first lamb I touch feels stiff and cold. Just a faint touch of warmth on its bluing tongue, I lower it despondently into the trailer. The next two, on an older ewe that has tried to get them up and licked dry, have some life in them but are fading fast, their core temperature dropping. Desperate measures are needed. I do what I have never before done and decide to save the lambs quickly and worry about the ewes later. If I have to catch all the mothers, these lambs will be dead. I will lose too much time. Two minutes later, I have gathered up five lambs and am on the road home.

Another ewe has lambed under a wall and had two proper strong lambs with big bold heads and white ear tips visible even in the mud. With a full trailer, I have to leave them to their mother's attention, but she is an old experienced ewe and knows the game. I meet a friend coming the other way from his own flock. We exchange blasphemies about who has experienced the worse mess.

Minutes later, I have the lambs tight under a heat lamp, hung so low it is burning off the slime, mud and afterbirth. I haven't much hope for any of them. The first one is stiffening like a corpse. There doesn't seem to be much to lose, so I stomach-tube it with some warm artificial colostrum, figuring something warm inside might help. But I may kill the lambs, because the shock of the milk is sometimes too much for them – I am gambling. I leave Helen drying them with towels from the bathroom. The children get themselves ready for school. Chaos. I go back for the mothers.

The fields are so sodden I am on my backside more often than I am on my feet. Only the bravery of Floss lets me catch them. All are hellish to catch because with no lamb to hold their attention they are free to gallop off. I fill the trailer with the required ewes (making a mental note which lambs they have each given birth to). Telling which ewes have lambed is made easier because they usually have a bit of blood or afterbirth on their tail, and they will usually hold to the place where they gave birth. Some of the ewes use their last bit of energy trying to evade being caught and look fairly weather-beaten and worn out.

I go back to the barn where Helen has

managed to get some life into the lambs, and an hour later, miraculously, they are all sitting up and warm. Each of them is penned with its mother, bedded with clean straw, and still with a heat lamp on it. The one that was in the beck is suckling its mother. By the time we have tended to them and had a bit of breakfast, shoved the kids on the school bus, wearing the wrong clothes, it is time to get back to the first lambing field to do the rounds again. The first time round in the morning is mostly about feeding the ewes, identifying any problems that can't wait, and then getting to the next flock. An hour or so later, we go round again and sort out the less urgent issues. But some days the troubles snowball, one damn thing after another into a furious blur of activity that makes a day feel like a week in normal time, with the stress mounting with each successive hold-up.

It can take one of us all day to look after the problem cases in the barn. If it can go wrong, it will go wrong at lambing time (imagine a couple of adults looking after several hundred new-born babies and toddlers in a large park).

I see an old ewe hanging determinedly to a sheltered place behind a hillock. When I get close, she stands and her waters break. She is not paining and is in good health. So I know it is fairly safe to leave her for maybe an hour to lamb naturally, then I will return and check all is well. I am now juggling in my head a bunch of things that need doing, being pulled in different directions.

I have a mental map of the sheep lambing at different places, and when I need to check again on each of them. It is like having a series of egg-timers in my brain for a number of ewes around the farm at different stages of giving birth. If this ewe hasn't lambed in an hour's time, I will catch her and see if there is a problem. I see another has had twins. I catch them and hold them up by their front legs so I can see how full their stomachs are. They are swollen, full of milk and warm. I don't need to worry about these lambs apart from a quick check later in the day. A few hundred yards away, I see another younger ewe stand up and a lamb fall out of her. She turns and licks it, mothering well. I can leave her for an hour or so to get suckled naturally. Then I need to check it.

The egg-timers in my head are always trickling away, reminding me of things I need to return to. Knowing when it is best to interfere and when it is not takes years of experience. My grandfather and father taught me that we have a range of options and the trick is to know which one to resort to, depending on the situation. A wild or stressed-out ewe might be best left alone, so that you don't make things worse. You can do more harm than good, they say, unsettling the ewes. My grandfather had incredible patience with the lambing ewes, and would leave them and leave them as long as all seemed well. He'd stand and watch, leaning on his crook, seeming to know when it was better to act, or when to leave well alone. I'd stand with him, wondering if he was right, and whether we shouldn't just catch the ewe and help it.

As I pass along the road between the fields, I see my neighbour Jean and she asks how the fell ewes I bought from her are doing. I call the flock I have bred for the last twelve years on our own land the 'beauty queens', and the new flock bought from her the 'fell flock'.

I suspect I was on trial with Jean for a few years to see if I was 'fit to look after her fell flock right'. She'd put thirty or more years into those sheep and she'd be damned if she was going to pass them on to some clown that would waste them. She knows enough about me to know I'm not a clown, but I'm also not a born-and-bred fell shepherd either, so I've been on probation. We spoke about this for two or three years, but then we needed to agree a price and terms.

I am in her kitchen, and the negotiations are to be done. Such haggling is part of the game, as long as it is respectful. She opens up with a cup of tea and some of her famous gingerbread; then a sermon she has clearly been thinking about for some time. She tells me how few flocks there are to match hers; that when she bought them she had to pay over the odds to the previous owner, Arthur Weir, because he knew they were good. She reminds me that it was when I was 'in nappies'. He had no doubt scrutinized her for years before doing that deal. She makes clear that these sheep are precious to her and this is more than simply a sale.

They are sheep that show the effort several generations of shepherds have put into them. Each autumn for centuries, someone has added to their quality with the addition of new tups from other noted flocks. There is a depth of 'good blood' in them. They are big strong ewes, with lots of bone, good thick bodies and bold white heads and legs. They return from that fell each autumn with a fine crop of lambs that are a match for most other flocks in the Lake District. She tells me that she had to pay £20 a head more for their being hefted, a charge that is carried on to each successive shepherd (like a fee for the work done in keeping them well hefted, separate from the value of the sheep). I will have to pay that now. She lists the prices of good fell sheep sold at auction in recent years by her and others, and tells me that they were 'drafts', not the 'stock ewes' that remained on the fell. She tells me that she has sorted out the old ewes, and is only selling young sheep with long lives ahead of them. Her 'good stuff' is still in the flock, and I will have to pay handsomely for it, that I am buying youth and quality.

I know all this is true. It is my move. I tell her that what she is asking for them is too much, that I am young and not rich, that I have two jobs, three kids and a mortgage. I already have a good flock of Herdwick sheep on our land that were once hers, but which have more than a decade of my breeding decisions in them. I can manage without hers. They are good, but no better than my own. I tell her the prices of most of the draft ewes in the auction over the past autumn, which was about a half of what she wants for these.

I tell her that no one else will pay the price she's asking for. I make it clear I respect her work and her sheep, but that the asked-for price is too high. She needs to come down.

She tells me that they are 'an investment'. That they will look after me for many years to come, providing me with an income and producing good offspring and draft ewes to sell because of the quality of blood in them. I am paying for something a long time in the making that will in turn last me a long time. She tells me that when she started out with her husband, they were hard up too, had to stretch themselves, had to work hard, but that over time these things came good. She tells me that by selling a few good tups from these ewes, I will have paid off much of their value. She softens a little on the asking price, but looks a bit wounded by the compromise.

The lower offer hangs in the air.

I decide to let it.

I sip my tea, trying to look disinterested. Perhaps it is working as she seems a bit defensive. After a while, I suggest an offer about £30 per sheep lower than hers, tell her I can't see them making more than that in an auction. But behind my mask I'm genuinely not sure; good stock ewes from respected flocks are rarely sold and can make good money. Still, I'm not being mugged on this deal if I can help it.

She now sits in silence, looking determined and tough.

No. No. No. The price I suggest will never do. It is robbery.

I soften and tell her that I will come upwards a little. I still think they are overpriced, but we have reached an impasse and someone needs to move. And I know that the chance to get a flock of sheep like this on your doorstep in the Lake District is rare indeed. I won't ever get another chance like this. This flock has to have a future as well as a past.

The afternoon goes by in a series of bids and counter-bids, interspersed with long broody silences. My tea cup gets refilled from time to time, but as the price drops, she stops refilling, as if an extra cup of tea is adding further cost to these painful negotiations. In the end, we agree a deal and shake on it. I've still no idea who came out on top, which is perhaps how it should be between folk that respect each other.

Of all the writers associated with the Lake District, Beatrix Potter (or Mrs Heelis, as she was known for her farming life here) is the one that I love the most. She had the utmost respect for the shepherds of the Lake District, and would have understood what took place between us in Jean's kitchen, because she, too, negotiated with shepherds to buy flocks of sheep.

When she bought her first true fell farm, at Troutbeck Park, she wisely asked the respected elder Herdwick breeders who might be a suitable shepherd. The name that emerged from these conversations was Tom Storey.

She went to see him and asked if he'd come and be her shepherd. He said he would if the money was right. She offered to double his money. He accepted. Later, she put him in charge of her flock at Hill Top Farm, near Sawrey.

You might think that her being Beatrix Potter, the famous and wealthy children's author, and owner of property, would have intimidated Tom Storey, a young Lake District shepherd; that her being notionally of a higher class and much older meant she would be due a degree of deference. You'd be wrong.

Soon after he became her shepherd, Tom fell out with Beatrix because she had got some sheep into the pens and 'redded' them for Keswick Show. She hadn't listened to him when he said they weren't good enough. He considered this to be ignorant meddling in his work. She protested that they had been show sheep in the past and tried to reason with him. He cut her off sharp. If she wanted to show those particular sheep, she had better get her old shepherd back. He would not show them and would leave. They weren't fit to show. As any shepherd knows, male or female, you're either in charge of the sheep or you're not.

She went back to the farmhouse and informed Tom's wife that he was bad-tempered.

Beatrix Potter could have got rid of Tom Storey when he defied her. But instead she worked with him, respected his knowledge and his beliefs, and learned a great deal.

In the years to come, they transformed the flock, and they would go on to win many shows. She knew it was his judgement that made a lot of this possible. She was very proud of their successes, and she, too, became respected for her knowledge of sheep. One of her best was a fine ewe called Water Lily, which won many prizes. In a classic old photo, Beatrix holds the prize in the background whilst Tom Storey holds the ewe proudly in the foreground. I won that same prize this last autumn.

Traditionally, this is the most un-English of societies. There is still, amongst us, a rough northern form of egalitarianism not unlike that which exists in Scandinavia. In Sweden they call it *Jantelagen* (the unwritten rule that forbids anyone to feel or act superior to his or her neighbour). Shepherds consider themselves the equals of anyone. The social status, wealth or fame of Mrs Heelis counted for little with Tom. They were, for all practical purposes, equals, and he was the superior party in many ways because of his specialist knowledge. When she worked on the farm (her property), she took orders from him for years.

Mrs Heelis hired an extra shepherd for lambing time each year to help Tom Storey. His name was Joseph Moscrop. He first went to help in 1926 and must have made a good impression because he was invited back every year for the next seventeen years. They wrote each other letters full of affection and friendship and mutual respect, the last one sent to him just nine days before her death in 1943.

I love these letters. Ostensibly they were written in order to agree the wage for Joseph's work, about which the haggling goes on for a long time each year, but because they are letters between friends, Beatrix also writes to him of the day-to-day, week-to-week happenings of a fell farm. Sheepdogs being too rough, waging war on the flies that make sheep suffer, the price of fat lambs, how the pet lambs are doing, whether the cattle are 'good doers', the merits or flaws of other sheep men, or whether Joseph knows of a good sheepdog for sale. She writes of sheep caught behind walls by snowdrifts, the falling price of wool, maggots on the sheep, shepherds being called up to fight in the war, and the state of the potato crop.

Lots of the names mentioned in those letters are those of the grandfathers of the men I know now.

Mrs Beatrix Heelis died on 22 December 1943. Her death was reported in the Herdwick Sheep Breeders' Association Flock Book in amongst the other respected members of the breed community who had passed away, a tradition that continues to this day. No more, and no less, important than the others. She would have asked for no more.

Her will is a remarkable document for someone who will always be known for her

311

children's books. It is not really about the books, but is instead full of concern for her legacy of farms, the ongoing care and respect of her tenants and the future of the fell-farming way of life. She put her money where her mouth was, handing fifteen farms and 4,000 acres to the National Trust. She stipulated that her fell farms should have fell-going flocks of the 'pure Herdwick breed'.

Soon after her death, her husband, William Heelis, the local land agent, sent a letter to Joseph Moscrop, asking him to come for the lambing. Joseph replied as he always had, demanding a higher wage, and Mr Heelis, who didn't know about the friendly haggling game between friends that Beatrix and Joseph had played through these letters for years (and being a little more formal in his social attitudes), replied to say that the suggested wages were impossible and that they should break the 'old connection'. So neither Beatrix nor Joseph saw lambing time at Troutbeck Park again.

The first year my new flock of fell ewes came down to the lower ground of our farm, they sulked all winter. They knew this wasn't their home. They'd stand in the corner of the field nearest to their old home as if in silent protest. They held to that corner even when there was better grass away across the field. They even held to that corner when the

312

snow came, and their stubbornness and clinging made fools of them because they were in the most exposed area of the farm. I had to walk them down from their place of protest to somewhere more sheltered. At lambing time, any that I had to bring into the barns with my own sheep were cross, and jumped out of pens, and hung to each other.

A lamb has gone missing. Its mother is agitated. She runs up and down the fence. I left them, hours ago, safe and the lamb well-mothered, and now it is gone. There are no clues. I ride around the field, checking the other mothers haven't stolen it or taken it by mistake. They haven't. I check the becks in case it has fallen in and drowned. We try to keep ewes with young lambs away from the becks, but it isn't always possible. I hate losing a healthy lamb. I check the neighbouring fields. No sign. Then I see that it has got itself stuck between the trunks of an old thorn tree, about a foot off the ground. It is fine, just squashed and tired. I lift it out and it runs off to suckle its mother.

You can lose hours looking for a lamb. Experience tells us that if you lose one lamb in a field, it can be an accident, but if you lose two or more then in all likelihood something is taking them. We have seen it and experienced it countless times. We have some fields that are well fenced and have no streams. If a lamb disappears, it has to have gone somewhere, and most lambs are too big for any bird to have taken.

We live with foxes skulking around the lambing fields overnight and in the twilight. Usually they are after the easy meal of the afterbirth left by the ewes. But maybe every other year we get a fox that starts to take new-born lambs, or even those a day or two old. Two years ago, we had a fox that was so bold it would be in the lambing fields in daylight when we were less than half a mile away, sniffing nearer to the new lambs and nipping in to grab the afterbirth, or a lamb's leg to snatch it away before the ewe could defend it. The older ewes are fierce, and stamp their feet and lower their heads as if to charge. But a younger ewe can be confused by the fox and can be fooled. Traditionally, the farms here called the local hunt when they had a rogue fox and it would come and get the culprit (or any other fox unlucky enough to be abroad when they came). The hunts have often found lamb bones, skin and remains near or in the fox den. Often the culprit is a bitch that has lost her mate, and has to become cunning to provide for herself or her cubs.

My elder daughter, Molly, is coming across the field with two ewes and their day-old lambs walking in front of her. She understands sheep, and cuts left or right behind them to keep them walking in the right direction. She knows when to pause and let them mother up again, because her grandmother has schooled her on this. I open the gate and the ewes lead the lambs through to the fresh pasture. My daughter is clutching her crook and beaming. She loves moving the lambs.

The mothering instinct is sometimes so strong that ewes will start to steal other newborn lambs for a day or two before they give birth themselves. They will appear behind a birthing ewe and will lick the lamb as it comes out, and nuzzle it away from its increasingly distressed mother. Sometimes we have to catch and pen a repeat offender to stop the chaos. Once she has her own lambs, she will be fine. Sometimes twin lambs will head away from their mother in different directions, despite her best efforts to nudge them back together; after a while they can be at either end of the field and she may not mother one when they are reunited. Sometimes several ewes give birth in the same place and then the lambs sliver and slide into each other, and I will scratch my head trying to work out which belongs to which.

The best way to prevent freshly lambed ewes getting muddled up, or trying to claim older ones that aren't theirs, is to clear the lambing fields of lambs each day. So we carefully move ewes with lambs to new fields with a bite of new grass grown for this purpose. It means that the ewes are given a boost of fresh grass when they start milking. The twins we often catch and load into the trailer behind the quad bike; singles can be walked away. Some ewes with strong maternal instincts will be caught easily. We exploit that mothering instinct to move them, with the ewe following us to a new field, or dry patch of ground, or to the trailer. Others are caught with the crook, or with the help of the sheepdogs (though only an experienced dog would be allowed anywhere near the lambing fields, as control is essential).

There is an invisible link between a ewe and her lambs; take liberties, or take your eye off the ewe for a second, and you can lose her, the link broken. Some ewes can abandon a lamb if they are stressed, so we are on a knife-edge helping them, and occasionally have to deal with a ewe that abandons or turns on her own lambs. It becomes a blur of adrenalin and stress, and though it wrecks you, it also comes and goes with a strange kind of buzz.

My daughter has gone for another ewe down the field. She lifts the lambs to their feet because they were sunbathing and sleepy. The ewe is a proud mother and stands over them, giving mouth. She is so defiant a mother that she nudges at my daughter with her head, pushing her away from her lambs.

Molly is having none of it and waves her away determinedly. She then walks the ewe, with lambs in train, up the field to me and I let them through the gate. My daughter laughs at the bossy ewe that was not remotely scared of her.

As lambs are born and straighten out in the first few days, we see the future of our flock, and little clues in their appearance tell us whether these will be good sheep. When I was fortunate enough to breed my best tup, Darwin, I knew soon after his birth that he was special. He was just brighter, stood better, and showed off more than the other lambs. He stood up with great legs, a big well-bred head and nice white lugs coming off his head at just the right angle. When I see one like that, I start dreaming about its future. I'm still waiting for another like him.

Lambing time often feels like it starts in the depths of winter and ends in summer. About halfway through, the spring comes and everything becomes easier. The seasonal transition is dramatic. The days are lengthening. The sun grows in warmth, hanging higher in the sky each day. The sheep have started to thrive again and put on flesh. The land is drying out. I can hear water seeping imperceptibly away. As the fields warm in the sunlight each morning, the valley is bottomed with mist and the fields are rimed wet with dew. The fell sides catch the warmth sometimes as it rises from the colder valley floor. This sun-warmed air catches my face as I rise up the fell sides to feed the ewes.

This morning I noticed something missing. The winter-vagrant fieldfares are gone from the hedgerows, back to the far North. I don't always notice straight away, and then I know I haven't seen them for a while. And you feel a little sadness, like the valley is emptier, quieter, less colourful and chatty. The fieldfares have headed away over the seas and other lands to breed. The fields are littered with fresh mole heaps, a rich black loamy soil of fields not ploughed for decades, if ever. The few moles we catch, we hang on the barbed-wire fences, making but a small dent in their number.

Then, as we work, we see the summer migrants return. Stonechats suddenly reappear from Africa, bobbing eagerly up and down on our walls, and on the bare ground where the ewes have been fed through the winter. Oystercatchers strut around the pastures and stand atop the gate stoops or posts. Curlews rise and fall on their own song. Geese circle past the fell side before falling slowly on muscular wings to the freshening pastures below. The trees host a little orchestra of whistle and chatter from the starlings. But the snow still clings on the high fells; Dodd Scar holds the snow sometimes right into May. The snow-speckled fell looks more and more like the side of a piebald pony. The buzzards start to find thermals and get back their majesty after a winter of sulking in the ash trees and scratching worms out of the molehills. Everything in the natural world around me changes and I am outside to feel and see it. Our spirits rise again.

I take off my waterproof leggings and wellington boots and throw them in a corner. I lace up my walking boots. I am back on top of the ground instead of wading through it. New grass means we stop feeding hay, or buying supplementary feed. Grass means our workload eases, or changes at least, to less life-critical tasks. Longer days mean we can start to do other jobs outside that have been on hold through the dark months. There is a time each spring when I know it can't go back, we are through it. I notice the oak buds fattening slightly. The catkins appear on the willow by the edge of the becks. And as we are lambing, the rooks begin their courtships and build their nests, journeying here and there with sticks, or tugging wool from the backs of our ewes to line their nests, leaving little circles where the robberies have taken place. They glide down from the wooded fell sides to where we feed without a single beat of their fingered wings, wheeling above us, seemingly motionless and within touching distance.

There is a kind of light-headedness that comes with spring. The whole valley echoes to the sound of ewes calling their lambs, and the older lambs start racing each other across the hillsides. Our work shifts from supervising lambing to looking after hundreds of young lambs and keeping them alive and out of trouble. Some days it all goes like a dream. Ewes give birth by themselves, get their lambs sucked, and then tuck them sheltered behind some rushes. Older lambs follow their mothers obediently and safely. Sometimes a lamb makes me laugh by chasing or trying to head-butt one of the crows that struts about amongst them, shining purple, bronze and jet black in the sunshine. I see frogspawn in the wet bits of land where I pass each morning. The heron folds down the wind heading downstream.

In the middle of April I take my geld hoggs (ewe lambs from the previous year that do not lamb) back to the fell. They have been away wintering in the lowlands on my friend's dairy farm. They are fat and bouncing around the fields in rude health. For a month, they have become increasingly restless and terrorize his walls and fences. They are bored teenagers and just when I least need the hassle (because I am lambing the ewes), I start getting phone calls to say they have escaped into a garden or someone else's fields. It should be a ten-minute job to gather them into the yard to give them their vaccinations before going to the fell, but it takes me half a day because they have managed to disperse themselves across the parish where he farms. They gallop about at 100 mph and test the dogs to the limits. Two of them are lame and decide to sulk and lie down in a bog. I carry them out on my shoulders. It is a great relief to lead them home in the trailer and shove them back up the fell where they can do no harm and be as wild as they like.

When I was a child, towards the end of lambing time the men circled the woods and shot crows, shouting, excited like boys, sociable again after the testing weeks. The whole valley echoed with cawing and the thuds of cartridges, twelve-bore retribution for one-eyed lambs and maimed corpses. Shattered twigs blown skywards, as nest floors crackled back down through the branches. Ravens, rooks, dopes (carrion crows), magpies and jackdaws ... all wanted for murder or GBH. Anything with black feathers was a 'dope'- a robber, a killer and a cheat. In a valley where men lived for their sheep, these shadows of the lambing field were guilty. The morning after, there were black specks in the rushes by the wood's edge: crumpled wings, perforated flight feathers, specks of blood, porcelain legs twisted and broken like cocktail sticks. The angry caws of the survivors reprimanded the valley and its shepherds.

A dope rose and twisted like a broken kite, before crumpling to earth like a bi-plane with torn canvas wings. 'Look at that murdering bugger, hoppin'' about,' Granddad said. A length of hazel, skin shining and soft to the touch, was sitting idle by the Land-Rover gearstick. My staff. My chance. We exchanged smiles that we both understood. 'You'll be wet to the skin,' he warned. But it was said as encouragement. I tore off across the pasture through a scattered lake of puddles, racing with the clouds reflecting beneath my feet. The crow rising and falling in desperate leaps to the sky. Wailing, thrashing wings, only one of which now held the air, the other failing, a deep black shimmering force like the wrath of God, and with one wing it still had me beat. But, legs pumping like pistons, I got almost within stick-reach. I caught the crow with the staff somewhere above its failing wings. It dropped like a toy into the standing water, suddenly small, and ended. I held it up in my hand, a trophy, and turned back to the gate, stick aloft. The old man didn't mind at all that I was soaked to the skin.

As lambing time peters away, and with plenty of grass and warmer weather, our focus changes to working on the ewes and lambs and getting them to the higher ground so that the meadows can be cleared for growing our hay. Last year's ewe lambs return from their wintering grounds and are driven to their home on the fells, holding there because that's what their mother taught them. In the valley bottom we have hundreds of sheep on the farm, each of the breeding ewes having either a single lamb or twins, and we need to begin the work of sorting them.

The crossbred lambs that will be grown and fattened for meat are castrated and have their tails docked (tails or testicles were cut off with a knife, or twisted off, with a spray of blood when I was a kid, but now it's done with an orange rubber ring that slowly stops the circulation and lets them fall off). Lowland sheep need their tails docked to stop them getting fly strike (maggots) on the damp and dirt that accumulates on their tails. The castrated lambs lie down for a few minutes, wincing, and then go off to find their mothers and seem fast to forget about it.

The fell lambs keep their tails because mountain sheep need them in bad weather, and can keep them because they don't scour on the fell-grazing or have mucky backsides that attract flies. The lambs all have to be 'doctored'; injected with vaccines for pre-ventable illnesses, wormed with an oral drench (as the parasites come to life again with the warmth), marked, tagged with their two fourteen-digit micro-chipped tags (a legal requirement), and notched in their ears to show they belong to our farm. We use a spray to stop blowflies laying eggs on our sheep, as in June and July some would be 'struck' without this and would eventually die a horrible death.

So in early May we gather the sheep into the pens or the barns. It is work we have always done as a family, because many hands make light work, although there is a clear pecking order amongst us.

The oldest members of the family always claim a kind of authority based on their experience, as I will when it is my turn. So my father reckons he is the boss, even though he is weaker after months of chemotherapy. Those of us in our prime do the catching and rougher work and work our dogs to get the sheep in and out of the pens. My mother gets the job of writing in the messy old textbook the individual tag numbers of each of the lambs so that we can always trace their ancestry. The children hold orange rubber castration rings, or the ear tags, and enjoy passing them to us. It is a day of trying not to forget what each ewe and lamb needs, with everyone given a task. But it is also a chance to identify the breeding decisions that have worked, and which haven't, so there is a lot to talk about. We can now see the quality of our lamb crop and which ewes have bred well and which have not. We spend the day forming judgements. Everyone has an opinion. Tups bought the autumn before are now cheerfully damned.

'That one's far too white out of that old ewe.'
'No it'll be alright... You wait and see.'

My grandfather would tell you where each lamb was born, or how it was bred: 'This one was born under that Scots pine at top of the horse pasture... I thought it was dead... But look at it now.'
These days of stories and chatting often threaten to take attention away from the confusing number of jobs each sheep needs done. My father will periodically shout, 'Oh fucking hell... We better stop talking, I've let that bugger off without marking it.' But the error will be corrected, the lamb caught and marked. Then this is followed by ten minutes of focused work, before the chattering cycle will be repeated.
We have some tremendous family rows on such days, but we get over it. When all is said and done, we are working as a family, and that is a special thing. My children now have opinions, and want to tell everyone how their sheep have bred, and how they are going to

beat their dad with a tup lamb when it grows up. My son is two years old and hangs over the rails, waving a crook, shouting instructions and offering unwanted advice. My dad smiles as if to say, 'Here we go again.' Nothing changes. Grandfathers. Fathers. Sons.

Once the ewes and lambs are 'doctored', they can be moved to the higher ground. As we push them out, the valley echoes with ewes and lambs calling for one another, all of them carrying our blue and red smit marks on their shoulders, as dictated since time immemorial in the Lake District Flock Book.

Beatrix Potter wrote about the smit marks on sheep in *The Tale of Mrs. Tiggy-Winkle*. She has the little girl asking the washerwoman hedgehog about the lambs' coats she is washing and is told:

'Oh yes, if you please'm; look at the sheep-mark on the shoulder. And here's one marked for Gatesgarth, and three that come from Littletown. They're always marked at washing!' said Mrs. Tiggy-Winkle.

Beatrix Potter knew the three farms around the Newlands Valley where the tale is set: Skelghyl, Littletown and Gatesgarth. She knew their flocks, their 'smit marks', and would have known of their shepherds. More than a century after she wrote those words, we still know those flocks, but God only knows what they think of these references in Japan, where her books sell so well.

When the fell flock I bought from Jean were walked back to their fell from our farm after their first winter, they remembered where they were going and streamed away at a trot to their summer home. The whole fell side seemed to sigh with relief at their being home again. The next winter they were fine on our farm, too, as though they had adjusted to the change.

Wordsworth may have 'wandered lonely as a cloud', but shepherds are social animals once winter is through. We come back together at the spring sheep shows in May, where we show the best of our tups (or ewes as well in the Swaledale events).

Hundreds of Swaledale shepherds congregate on a windy bit of moorland next to Tan Hill Inn. The tiny road that winds over the moor is shadowed with parked cars for half a mile or so on either side of the show field, an area shaped out of wooden hurdles that springs up overnight and vanishes as fast afterwards. It is one of the great achievements of that breed to 'win Tan Hill'. But truth be known, it is also an old-fashioned coming together. I understand that, in ancient times, places like ours were governed at gatherings like this, big fairs where people came together, showed or traded livestock, raced horses, got drunk and made friends or found wives or husbands. At Tan Hill, shepherds chat with their friends from across the Swaledale lands that they have not seen since the autumn sales, comparing notes about their lambing and how well the tups have done. There are other smaller spring fairs like this across the fell country. Without these gatherings we would drift into being strangers divided up into the different valleys, and our breed communities would drift apart.

The Herdwick shepherds have their own spring tup fairs. Keswick Tup Fair takes place on 'the Thursday after the third Wednesday in May' (work that one out) on Keswick Town Field. The other tup fair takes place a week earlier in Eskdale in the field by the Woolpack Inn. Folk gather each spring from right across the Lake District, coming to return their hired tups and to show their best ones against each other, as they have for centuries. The field is divided by wooden or metal hurdles into a big circle of bale-string-tied pens, a makeshift ring in the centre. Sheep are unloaded from aluminium sheep trailers towed behind Land-Rovers, and walked to pens set aside for each farmer. The tups blink as they come out into the daylight, and strut as they realize they are amongst strangers, giving an occasional thump of heads and horns.

The sheep are in full wool, as befits the time of year; some might even be scraggy, having lost or rubbed off some of their fleece – but this is ignored because it's irrelevant. At the western corner of the field is a large pen where the tup hoggs (last year's male lambs) are penned together, their dark chocolate fleeces contrasting with their faces. They are penned together because one of the most enjoyable bits of the day is the open judging competition with everyone seeking to pick out of this mass of thirty or forty hopefuls the real stars. These youngsters are a year old, but in Herdwick terms they are mere babies, unproven, and not worth much until they mature out without faults. The best will go on to be stock tups and wield an influence on the breed for years to come. It is like sorting through the junk at a jumble sale looking for a Rembrandt. They are powerful and fit after wintering away in the lowlands, and are pawed over for any faults: a 'twined' leg (slightly twisted); 'bad mouth' (teeth twisted or gapped, or over-shooting the top jaw rather than nestling on its pad); a poor or 'plain' skin (a fleece too open and soft for this landscape). Some might dip behind the shoulder, or have a fleece that suggests it will go too white and lose its colour in the next year. The line between hero and zero is minuscule. It might take a person ten years to learn to see these things, and even then be little more than a half-educated amateur.

When I visit my shepherding friends, the walls of their houses, and their mantelpieces, are like shrines to the best tups they have bred. In the Herdwick breed, the most noted flock in recent years has been the Turner Hall flock from the Duddon Valley. Turner Hall was built by folk who intended staying put, built to last. The stone-built farmhouse and barns are tucked amongst the rocks and the trees of its rugged stony valley. But each autumn, out of those humble fields, come some of the best Herdwick sheep. The farmer, Anthony Hartley, knows more than I ever will about Herdwick sheep. I pick his brains endlessly, hoping to catch up in knowledge someday. For me, his sheep are the benchmark. Several generations of Hartleys have made it their business to breed great Herdwick sheep. Look at the old black and white photos and someone in them is usually a Hartley, often with a thoughtful expression on his face as if he is thinking something through.

In the old barns at Gatesgarth, where another of our friends, Willie Richardson, farms, the beams are adorned with a century of rosettes and prize certificates for great Herdwicks, some tattered and disintegrating, with fading colours, others more recent.

Some of them date from when Mrs Heelis would have been showing her sheep against Gatesgarth sheep. In places like those old stone barns is the real history and culture of the Lake District. But only the shepherds and shepherdesses working on the sheep ever see them; thousands of visitors pass by the barns on their walks and never know they exist.

Town Field in Keswick and its illustrious Herdwick tup fairs go unmentioned in most books about the Lake District, but this modest and unheralded patch of grass by the River Greta is sacred soil, a theatre of shepherding ambitions, and perhaps the most important meeting place for the Herdwick breed. The historic function of the fair is to return the tups hired the autumn before to their owners, having been 'wintered' on another farm: it is a way to spread the bloodlines around the valleys and get young tups grown out at no cost to the owner by earning their keep elsewhere. But now it is spring they should be back on their own farms, so they can be shoved up the intakes to the coming grass, and later in the year made ready for the sales. I have never even come close to winning Keswick Tup Fair, or the Edmondson Cup. Someday I'll win it, or at least die trying.

My mother says we get 'tup fever'. A kind of

insanity takes over, starting in the spring and building to fever pitch by the autumn, until the shows and sales become all we think about. She could be right. Suddenly, one evening in late spring or early summer, some sheep-breeding friends will call round, professing to being just out for a 'ride out'. But they haven't really come to socialize. They have come to have an early look at the tups, and to check out whether our lambs look as if they will be good in the shows. A proud shepherd never wants anyone to see their sheep when they are not at their peak. We suffer each other's nosiness and play all sorts of games, hiding the best in fields far away from the road and prying eyes; pretending to show people our best; keeping the stars hidden until it matters.

There is great skill in the preparation of these special sheep for the shows and sales. Herdwick tups (and the best ewes) are not sold with their fleeces in their natural slatey blue-grey colour, but are, as they have been since time immemorial, 'redded'. No one knows for sure why it is done or when it started. It is just done and always has been. There are two theories: first, that shepherds some centuries ago wanted to see their most valuable sheep on the fell sides with ease, so they coloured the tups with the brightest natural colour they could find; or, secondly, that this is some ancient form of animism, that people here might have worshipped their sheep in some way as far back as Celtic times, and coloured them as some form of ritual. Knowing the way people here think about sheep to this day, I find the second explanation very easy to believe.

The palms of my hands are red as if they have been soaked in the blood of a mountain. The raddle has a deep iron-ore tinge to it. Once, the colouring would have been taken from the rusting rock faces, the brightest natural colour they could find. In front of me is a Herdwick tup, bristling in its blue-grey coat. He is held by my father. He bridles as I step towards him, and I see my father's knuckles whiten as he takes hold tighter. I place my redded hands at the base of his neck where the grey of his mane starts. I pull my redded hands back along the wool of his back. The paste leaves a track of colour along his back. I push and pull my palms back and forth on his back until there is a two-hand width of raddle.

All of the traditional breeds of sheep have these strange ceremonies. The red changes a Herdwick sheep, helps the contrast between the fleece and the snowy head and legs. When we wash their faces and legs the day before a show or sale, the sheep come up bright white, and they take on a noble and handsome appearance. They have transformed from their work clothes to their Sunday best. 'Herdwick Show red', a dark rusty-red powder, is now bought in a bucket. The Swaledale equivalent is to colour the fleece of the tups and ewes for sale in peat, often dug from some special secret location on the moor that has been shown to provide just the right tint of peat bog to meet the ideal of beauty required.

All thoughts through the spring and summer lead to the autumn, when everything the shepherds know is tested in the shows and sales, in the full glare of scrutiny and the judgement of their peers. This isn't just vanity, though there is vanity in it, and it isn't just pride, though you will never meet prouder folk. This is the coming together of everything, the ending of old stories and the beginning of new ones. The great flocks of sheep represent the accumulation of countless achievements at these shows and sales over many years. Each year's successes or failures layer up like chapters in an epic novel. The story of these flocks is known and created in the retelling by everyone else. Sheep are not just bought, they are judged and stored away in memories, pieces of a jigsaw of breeding that will come good or go bad over time. Our standing, our status, and our worth as men and women, is decided to a large extent by our ability to turn out our sheep in their prime, as great examples of the breed.

I once bought a little Herdwick shearling tup (in his second autumn) from Willie Richardson from Gatesgarth at the sale at Cockermouth. He was by popular consent agreed to be a beautiful, stylish little sheep, perfect white in his head and legs and where his legs met his body. He had only one fault, that he was probably a bit too small. So he cost me just £700; if he had stood a few inches taller he might have cost me another thousand pounds or so. I shared him with a young shepherd, but three weeks after the sale he decided we had made a mistake, that the tup was too small, so he never put any ewes to him. Before long I was being teased about this little tup, the consensus being that I was wrong, he would breed too small. I nearly listened to everyone pulling him apart, but something told me not to, so I gave him the best of my ewes the first autumn. A gamble. That was six or seven years ago, and now his daughters are, I believe, some of the best-looking and breeding ewes in the Lake District. That little sheep was one of the best we have ever had. He mated with just ten ewes last autumn, and then was found lying, old, worn out and dead, in the middle of the field. Some of the best shepherds, who once dismissed him, now admit, when I remind them, that they were wrong about him.

Sometimes these things work, sometimes they don't.

Bea climbs over the pens and quietly but determinedly takes my show lamb from my hands. We are at one of the shows we try to win each year. The judge, Stanley Jackson, comes along the line and smiles when he sees she is holding it tight around its neck. She is cute, so the other shepherds tease me and say it is just a way to sway the judge. I tell them to bugger off, that there is a new shepherd on the block and they'd better watch out. Away in the next set of pens, my father is showing his Swaledale sheep. My other daughter, Molly, is holding one of his and it wins its class. Three generations of us doing what we do. Other families are spread out like this around us. The lamb Molly is holding was sired by the tup that my father and I bought the year before, the one that he had admired on Christmas Day from the window, when I thought he was going to die. He has seen this little dream come true. He looks suntanned and happy. The cancer may still be inside him, and may someday have the final say, but for now he is alive, living a life he would not swap for all the riches in the world.

Summer starts when the last of the sheep have lambed, and the marked and vaccinated flock is driven up the valley sides, either to the allotments or intakes if they have twin lambs, or to the fells if they have single lambs.

Many fell farms are located at the bottom of the fell that they have grazing rights on, so it can be as simple as opening a gate and letting the ewes take their lambs on to the fell that starts the other side of the fence or wall. Other flocks like mine need to be walked miles to their head. A trickle of ewes and lambs will make their way up the sheep trods, paths worn by the sheep over the centuries, and slowly spread out across the mountain until they find the place where they belong. Their sense of belonging is so strong that some have been known to go straight back to where they were heafed with their mothers, an irresistible urge within them to head home to their 'stint', even if they haven't been to the mountain for three or four years.

Weeks later we are clipping a batch of ewes in our barn. They are Herdwick ewes that came down from the fell a few days before. These days I am a lot faster than my dad. I can do nearly two sheep to each one he does. That is as it should be, because he has reached retirement age, and I am in my clipping prime.

He knows that I'm still not as hardened to work as he is, and that if we kept going for hours I would slow down, that I'm not as fit as he was at my age. You can sense the person clipping next to you, sense when they are struggling or when they are flowing well.

He knows I am clipping as well as I ever have done before. For many years I struggled to match his speed, and I would get frustrated or angry at my lack of stamina or technique, so part of me enjoys letting him know that now, eventually, I can beat him as he once beat me. I give him a cocky smile from time to time as if to say, 'You once tortured me like this and now it is your turn.' He smiles awkwardly like you do when the speed isn't there and you are beat.

Then I notice him stand up after finishing a sheep. He walks quietly away, and I know that something is wrong. I ask if he is OK. He smiles as if to say everything is fine, but I know it isn't. He is feeling some pain, something that is robbing him of his strength. He lets me shear the final few sheep.

I am forty years old and I have never once seen my father take a step back from work. Not once. My dad is one of the hardest men I have ever met. I have known days when we worked like dogs and completed our own work, and I'd be dreaming of a hot bath or watching the TV, and then he'd realize that a neighbour was still working and might need a hand, and he would go and help them, and he'd volunteer me, with no sense of it being of any kind of benefit to us whatsoever. I'd ask what we were doing, when we had more than enough work of our own, and he'd pretend not to hear the question. Then, when the work was done, he'd drive me mad by waving away the suggestion that the neighbour might pay us.

It was like his code of honour. Work that needs doing should be done. Work is its own reward. Never step back from work or you look bad.

But something was different now. My dad just walked away from work. It is the most un-him thing I have ever seen him do. He straightens his back and walks away – and we both know that IT is in him again.

The days when we walk the sheep back to the fell are the best moments of my year. There is nothing like the feeling of freedom and space that you get when you are working with the flock and the dogs on the common land. I escape the nonsense that tries to consume me down below. My life has a purpose, an earthy, sensible meaning.

Gavin Bland, a friend of ours from the largest and thus perhaps one of the most important Herdwick farms of them all, West Head, summed it up when he told me recently that he could not farm a lowland farm now with small fields and fences everywhere: 'When you're used to big spaces and having no one around you, you get used to it. I couldn't be done with being fenced in among too many other folk.'

Ours is not such a large, or tough, fell farm as West Head, it is a 'cabbage patch' by comparison, but when I go to the fell, I know what he means. Once tasted, it would be hard to walk away from.

This is an ancient, hard-earned, local kind of freedom that was stolen from people elsewhere, the kind of freedom that the nineteenth-century 'peasant poet' John Clare wrote about. He lamented the changes in the Northamptonshire landscape he loved because of enclosure. He saw the disconnection that was being created between people like him and the land, something that has only got worse with each passing year since then. Across most of England over the past couple of centuries, common land has been enclosed until only islands of it have been left, in poor or mountainous places like ours where something older remains. Ours is a rooted and local kind of freedom tied to working common land – the freedom of the commoner, a community-based relationship with land. By remaining in a place, working on it and paying my dues, I am entitled to a share of its commonwealth.

Working up these mountains is as good as it gets, at least as long as you are not freezing or sodden (though even then you feel alive in ways that I don't in modern life behind glass). There is a thrill in the timelessness up there. I have always liked the feeling of carrying on something bigger than me, something that stretches back through other hands and other eyes into the depths of time. To work there is a humbling thing, the opposite of conquering a mountain, if you like; it liberates you from any illusion of self-importance. I am only one of the current graziers on our fell (and one of the smaller and more recently established ones at that), a small link in a very long chain. Perhaps, in a hundred years' time, no one will care that I owned the sheep that grazed part of these mountains. They won't know my name. But that doesn't matter. If they stand on that fell and do the things we do, they will owe me a tiny unspoken debt for once keeping part of it going, just as I owe all those that came before a debt for getting it this far.

When I leave my flock in the fells surrounded by grass and come down home, I leave something of myself up there with them. So I look away to the skyline where they graze several times a day. Sometimes I can't help myself, and go back up the fell just to see that all is well. The skylarks ascend, singing, disturbed by my boots and the sheepdogs.

The sheep's evident satisfaction at being back where they feel at home means that winter and spring are fast receding behind us. The fell sheep can largely look after themselves in the coming weeks. So I lie down by the beck and cusp out a handful of water. I slurp it. There is no water tastes so sweet and pure.

Then I roll over on my back and watch the clouds racing by. Floss lies in the beck, cooling off, and Tan nuzzles into my side, because he has never seen me lazing about. He has never seen me stop like this. He has never seen summer before.

I breathe in the cool mountain air. And watch a plane chalking a trail across the blue of the sky.

The ewes call to the lambs following them as they climb up the crags.

This is my life. I want no other.

Acknowledgements

It is humbling to discover when writing your first book how many people work hard to make it happen. The words and photos are mine, but a lot of other people are responsible for the book being in your hands. Thanks to all of you; I have loved writing this book.

Thanks to my agent Jim Gill, of United Agents, who sold this book before I had even met him. Jim believed in it and helped me find the right publisher. With a young family and other commitments, I needed an advance to be able to write this book, and Jim got it for me. I knew nothing about the world of publishing, so he guided me through that too. Thank you.

Thank you to the other editors who tried to buy the rights to the book; your interest and kind words encouraged me and reinforced my sense that it mattered and could work.

Huge thanks to Helen Conford, my editor at Penguin. Helen was willing to invest in me to write a book that was then still largely in my head. Helen believed in it, and from our first conversation I knew she was brave and respected what I was trying to do. I needed a great editor and I got one. Thanks also to Casiana Ionita, Stefan McGrath, and the rest of the brilliant team at Penguin.

Thanks to Colin Dickerman, James Melia, Marth Schwartz and the rest of the team at Flatiron Books.

Thanks to Julie Spencer for giving me opportunities to write and pushing me to do better.

Thanks to Alexis Madrigal and Robinson Meyers at *Atlantic Monthly*, who helped this book to happen by publishing an article in November 2013.

Thanks to Richard Eccles at *Cumbria Life* magazine for printing my monthly column and giving me the freedom to do things that helped me write this book.

Thanks to the more than 30,000 people who follow our farm on Twitter (@herdyshepherd1) and who have been hugely supportive and encouraging. I've learned loads from you. You might be surprised my name is on the book ... I clung to being anonymous for as long as I could get away with it! I have no interest in personal celebrity; our way of life is much more important than me.

A lot of people have helped me to try to understand the literary and artistic history of the Lake District – I thank them all. Particular thanks are due to the following people. Professor Angus Winchester, Lancaster University, a fine historian who has, in person and through his books, taught me a great deal about our landscape and its past. John Hodgson, Lake District National Park Authority, has been endlessly helpful and patient and has tried to help me to understand the evidence of the human history of the Lake District. Linda Lear's excellent biography of Beatrix Potter was an invaluable resource that helped me to write about her and her shepherds. I've also learned a great deal from the development of the Lake District's World Heritage nomination process over several years: thanks are due the members of the Technical Advisory Group 2. Even when I sometimes didn't agree with you, I was learning. My debates with Ian Brodie at Lancaster University helped me sharpen my ideas and understand better different perspectives on the Lake District. Thanks also to Julia Aglionby, who knows a lot more than me about the legal complexities of common land and who kindly tried to share her wisdom. Thanks to Michael McGregor, Geoff Cowton and the late Robert Woof at the Wordsworth Trust – who all helped me better understand Wordsworth. My friend Terry McCormick was crucial in helping me understand the writings of Wordsworth about farming and shepherds. Eric Robson was a great support and shared ideas and Wainwright anecdotes with me (he tells me Wainwright was fascinated by the fell shepherds when they met).

Thanks also to William Humphries, Rose Dowling, Mike Clarke and Emma Redfern for reading the final draft and making comments.

Any flaws and inaccuracies that remain are mine and mine alone.

Thanks to all the excellent people I have been fortunate to meet around the world at World Heritage sites and through UNESCO, who have helped me to understand why stories and rooted identities matter, and what a 'cultural landscape' really means.

Heartfelt thanks go to Mrs Judith Craig of Morland Primary School, who helped me to love books and learning and who encouraged me later on from a distance.

This book tells the story of me, my dad and my grandfather – but truth be told I can't credit them with the book itself. They weren't book people. Instead, the women in our family deserve some credit...

Thanks for everything, Mum. You helped me to love books. And thanks for listening to me ramble on about books or ideas while we worked on the farm or while you were ironing or cooking etc. I'm sorry I put you through stuff. I know you are deeply private, so I am sorry if the book embarrasses you. I felt it had to be honest and open or it wouldn't work.

Thank you to my kids, Molly, Bea and Isaac, just for being you and being there (even when I needed peace and quiet and didn't get it ... it still got done). I don't care if you become farmers or not; I just hope this book helps you to understand us, and go into the world knowing who we are – proud. They can't take away the stuff in your head and your heart. Hold on to it.

Thanks to my wife, Helen, for everything. You have always been my best friend. You worked so hard with me on this book, and on everything else we do. It's been mad, a roller-coaster ride. Most people get beaten back by life, but your support helped me stubbornly hold on to the dreams I had. And someone has to pick up the pieces when I'm mentally AWOL. I could barely hold a pen when I met you, and I didn't really understand grammar or the rules, so thank you for being patient and putting up with me.

Thanks to the farming people I grew up amongst and who I am proud to call my friends. There are too many of you to name individually, but thank you one and all. This is my version of the story of my family, but there is nothing exceptional about us – we are just one of hundreds of such families. This is just one story, one perspective, amongst many. I hope the book helps other people to see what we all do, and show it greater respect in the future. I don't want to lose the amazing patchwork of family farms that make this landscape what it is, and I don't think many other people do either. Keep going.

Finally, thank you to my dad for everything. I hope this book makes clear how much I love and respect you. Keep fighting.

This Large Print Book, for people
who cannot read normal print,
is published under the auspices of

THE ULVERSCROFT FOUNDATION